D0716493

Learning Autodesk Inventor 2008
The 2D to 3D Transition Handbook

600068570

Learning Autodesk Inventor 2008
The 2D to 3D Transition Handbook

Ralph Grabowski

Autodesk

DELMAR
CENGAGE Learning

Australia • Canada • Mexico • Singapore • Spain • United Kingdom • United States

Autodesk

Learning Autodesk Inventor® 2008
The 2D to 3D Transition Handbook
Ralph Grabowski

Vice President, Technology and Trades SBU:
Dave Garza

Director of Learning Solutions:
Sandy Clark

Managing Editor:
Larry Main

Senior Acquisitions Editor:
Jim Gish

Marketing Director:
Debbie Yarnell

Channel Manager:
Kevin Rivenburg

Marketing Coordinator:
Mark Pierro

Production Director:
Patty Stephan

Content Project Manager:
Betsy Hough

Editorial Assistant:
Sarah Timm

Senior Project Manager:
John Fisher

Production Manager:
Stacy Masucci

Book Design and Typesetting:
Ralph Grabowski

Cover Images:
Cover imagery courtesy of Conveyor Lines LTD, United Kingdom Rendering by Stephen Hooper, Autodesk Inc. United Kingdom

COPYRIGHT © 2008 Autodesk, Inc. Autodesk, AutoCAD, Inventor, and the AutoCAD logo are registered trademarks of Autodesk. Delmar Learning. Thomson, the Star Logo, and Delmar Learning are trademarks used herein under license. Delmar Learning uses "Autodesk Press" with permission from Autodesk for certain purposes. Certain drawings copyright by Bill Fane and used by permission.

Printed in Canada
1 2 3 4 5 XX 11 10 09 08 07

For more information contact
Cengage Delmar Learning
Executive Woods, 5 Maxwell Drive
PO Box 8007
Clifton Park, NY 12065-8007

ALL RIGHTS RESERVED. No part of this work covered by the copyright hereon may be reproduced in any form or by any means — graphic, electronic, or mechanical, including photocopying, recording, taping, Web distribution, or information storage and retrieval systems — without the written permission of the publisher.

For permission to use material from the text or product, contact us through
Tel.   (800) 730-2214
Fax   (800) 730-2215
www.thomsonrights.com

Find us on the World Wide Web at www.delmarlearning.com

Library of Congress Cataloging-in-Publication Data:

ISBN-13: 978-1-4354-1329-0
ISBN-10: 1-4354-1329-6

THE SHEFFIELD COLLEGE
CASTLE COLLEGE

	600668570
Coutts	05.03.09
	31.99
	L

WITHDRAWN FROM STOCK
620.0042
DATE Aug '18 INITIALS

NOTICE TO THE READER

Publisher does not warrant or guarantee any of the products described herein or perform any independent analysis in connection with any of the product information contained herein. Publisher does not assume, and expressly disclaims, any obligation to obtain and include information other than that provided to it by the manufacturer.

The reader is expressly warned to consider and adopt all safety precautions that might be indicated by the activities herein and to avoid all potential hazards. By following the instructions contained herein, the reader willingly assumes all risks in connection with such instructions.

The publisher makes no representation or warranties of any kind, including but not limited to, the warranties of fitness for particular purpose or merchantability, nor are any such representations implied with respect to the material set forth herein, and the publisher takes no responsibility with respect to such material. The publisher shall not be liable for any special, consequential, or exemplary damages resulting, in whole or part, from the readers' use of, or reliance upon, this material.

Brief Contents

Table of Contents

Introduction

AutoCAD users feel comfortable with their software. Last night, I attended an CAD user group meeting. A survey of hands showed that most attendees used AutoCAD, and most of these used AutoCAD 2005. A few employed add-ons or made use of other CAD packages. Nearly all were hesitant about switching to a "real 3D" modeling software package, such as Autodesk's Inventor and Revit. The reason for the hesitation? They believe AutoCAD works adequately for them.

This book's technical editor, Bill Fane, was at the meeting too, and stayed late afterwards enthusiastically showing the rudimentary benefits of Inventor. Change a dimension on a 3D part, and the change ripples — *automatically!* — to all connected parts, as well as to the 2D drawings, parts tables, and so on. The AutoCAD users reacted excitedly, peppering Mr Fane with questions as they became aware of how much easier Inventor could be in their lines of work.

And this is the purpose of the book you are holding in your hands. It's to help AutoCAD users like you who may hesitate to jump into Inventor, that very different-seeming CAD package.

The Purpose of this Book

Learning Autodesk Inventor 2008 describes the similarities and differences between AutoCAD and Inventor. It reassures you that your valuable collection of drawings can be reused — even combined — in Inventor, yet be taken beyond the drawing-with-lines-and-arcs structure of AutoCAD.

Over the last several years, Autodesk has tweaked AutoCAD and Inventor to become more compatible with each other. The user interfaces are somewhat similar, numerous commands are complementary, and *.dwg* files can be reused in both directions.

Above all else, it is important to understand this one fundamental difference between AutoCAD and Inventor:

- In AutoCAD, objects drive dimensions.
- In Inventor, dimensions drive objects.

In this book, I'll work with you to recognize the impact this difference has on the way you create drawings. It boils down to this: it no longer matters what size things are initially. *__Just draw it__*, and worry about sizing it later. As Bill Fane puts it, "Inventor lets us get back to napkinCAD."

Which is why Inventor calls 2D drafting *sketching*. You are about to enter a brand-new CAD experience where you no longer need worry about text sizes, linetype scales, paper space, and plot styles. Instead, you sketch, and then you extrude, and later you combine; and then your drawing is done. Inventor takes care of nearly all the housekeeping details that ordinarily take up several chapters in how-to books. The way I figure it, for an extra thousand bucks or so, you get a CAD program that's going to make you *a lot* more productive.

This book assumes you are already familiar with 2D CAD concepts, and that you are an AutoCAD user. It dives right into illustrating the advantages of Inventor, and then describes those things that Inventor does differently from AutoCAD. Along the way, I throw in plenty of tutorials to help you develop an appetite for the Inventor Way of Doing CAD.

About the Author

Ralph Grabowski has been writing about computer-aided design since 1985, when he became Technical Editor of *CADalyst*, the original magazine for AutoCAD users. He is now editor of *upFront.eZine*, a weekly e-newsletter on the business of CAD, and the author of over one hundreds books about AutoCAD and other software. This is his first book on Inventor.

Mr Grabowski's Web site is at www.upfrontezine.com and his Weblog is at worldcadaccess.typepad.com.

Acknowledgments

First of all, I have **Robert Kros**s to thank for asking me to write this book. Mr. Kross is senior vice president of the Manufacturing Solutions Division at Autodesk. But then the book would never have happened without **Marci Lackovic**, marketing manager, on the MSD Community Marketing Team at Autodesk.

I thank my editing team of **Stephen Dunning** and **Bill Fane** for squeezing in editing among their other summer activities. Dr Dunning is professor of English literature at Trinity Western University, and copy edits my books with precision. After 22 years as a professional writer, I continue to learn from him about how to write better.

Mr Fane is instructor of mechanical technology at the British Columbia Institute of Technology, and he performs technical editing on my books with humor, some of which I capture in this text. He does stuff with Inventor that Autodesk programmers say isn't possible, yet has infinite patience for teaching newbees correct modeling technique. (Well, except when using his sergeant's voice on students cruising the Internet during his lectures.) Mr Fane kindly provided the tutorial models for chapters 4 and 6.

Additional technical editing was patiently provided by **Chris Palmatier**, Product Designer for the DWG TrueConnect project; **Simon Bosley**, Inventor Product Manager; **Grain Gardiner**, Inventor Technical Marketing Manager; and **Jay Tedeschi**, Senior Solutions Evangelist, all of Autodesk.

Thanks to my wife, **Heather**, for putting up with me churning out these 304 pages during the summer months. Perhaps next year I'll get to repaint the exterior of our home. *Soli deo gloria.*

Ralph Grabowski
28 September 2007
Abbotsford BC Canada

Part I

Inventor for AutoCAD Users

Chapter 1

AutoCAD User,
Meet Autodesk Inventor

Welcome to Autodesk Inventor!

Inventor is Autodesk's software for creating 3D mechanical models that simulate products in the real world. It's most commonly used to design machinery, and so Inventor can design steel frames, electrical wires, hoses and pipes, and gears and belts. And it completes these designs more quickly and conveniently than any top gun AutoCAD user could hope for — and just for $1300 more than AutoCAD.

In this chapter, I'll introduce you to some of the tasks Inventor is capable of, tell you about its hardware requirements, and brief you on the history of Inventor.

You may have heard of drafters who cheat in AutoCAD drawings by extending lines to make them fit, or placing unscaled dimensions. But there's no cheating with Inventor. That's because Inventor is a *history-based parametric solid modeler*. If it can be designed by Inventor, it can be built.

IN THIS CHAPTER

- 3D terms used by Inventor.
- The Inventor workflow.
- Design assistants and speciality models.
- The benefit of history-based design.
- Computer requirements for Inventor.
- Installing Inventor.
- Getting more assistance.
- A concise history of 3D at Autodesk.

INVENTOR 3D TERMS SUMMARY

Assemblies are complete 3D models, made of two or more parts.

Parts are single parts or models, such as a bolt or a handle.

Constraints connect parts connect to each other to form assemblies; they also keep lines and other 2D objects together in sketches.

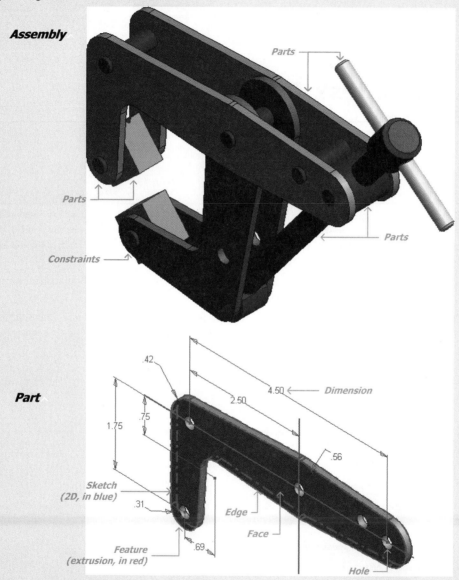

Assembly

Part

Features define parts, and are typically generated by extrusions, revolutions, and so on.

Sketches define features, and are usually drawn in 2D.

Dimensions define the size of sketches.

Holes are voids in parts, usually where bolts go.

"Solid modeler" means that Inventor creates digital objects that simulate real-world objects. The software computes the mass, appearance, and other properties of models based on assigned material types, such as steel, copper, or rubber.

"Parametric" means that objects are not fixed in size. Instead, their size depends upon dimensions and connections to other parts (called "constraints"). Sometimes you will hear the phrase **dimension-driven design**. This indicates that dimensions are determining the size of parts; change the dimension, and the part changes too. That's the opposite of AutoCAD, where the size of parts determines the values displayed by dimensions.

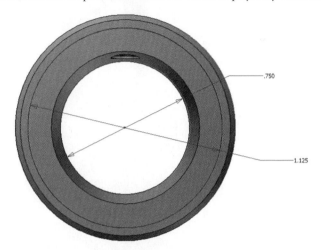

Inventor using dimensions to determine the size of parts.

The connections between and within parts are known as **constraints**. Constraints ensure two sides of a part are perpendicular or parallel to each other; they also ensure that two parts stay attached at specific points, and that the size of one part affects the size of another part, even when they are some distance apart.

The purpose of constraints is to capture these relationships (also known as design intent) so that the author and others can modify the model.

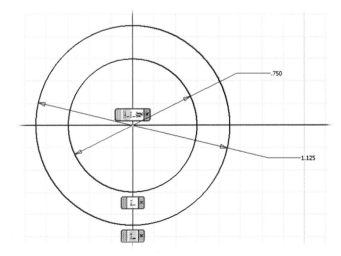

Constraints connecting sketched circles.

"History-based" means that Inventor tracks editing changes made to solid models. (A simple version of history is found in AutoCAD 2008; it shows source primitives of solid models.) As you add extrusions, fillets, holes, and other features, Inventor adds them to the history tree, which it calls the "Model browser." Changes made to objects later in the tree affect objects earlier in the tree, and vice versa. This forces you to employ a logical order of operation, which allows others easily to see how you arrived at the completed model.

Left: History tree (Model browser) listing parts used to make the padlock assembly.
Right: Red lines highlighting the selected part, a retainer.

Combining parametrics with history gives you an important advantage: making changes quickly. For instance, to generate hose connectors with diameters of 1", 1.5", and 2", you don't draw three hose connectors; you draw one, and then change its ID (inside diameter); Inventor updates the rest of the model based on the relationships defined by constraints.

The goal of Inventor is to make it easy for you to complete designs quickly by including:

- Styles and international standards.
- Pre-drawn content of industry standard parts.
- Vault software for finding and reusing designs.
- Design guides known as Design Accelerators.
- Engineering handbooks.
- And more.

You get to experience some of Inventor's ease-of-use during the quick look and design sessions in later chapters.

The Inventor Workflow

When drafters create 2D drawings with AutoCAD, they usually follow a regular set of steps. A typical workflow involves opening a template, and then laying down lines and other 2D elements. These are enhanced with hatch patterns and linetypes. The drawing then is opened in layout tab to place text and dimensions. The drawing border is added and the title block filled out. In the end, the AutoCAD drawing is plotted on paper or electronically.

Drawing with Inventor also follows a regular set of steps. The typical workflow also starts with using a template, and then laying down 2D *sketches*. These are extruded into *features* to create *parts*. The parts are combined into *assemblies*, and then the assemblies are opened in the Drawing environment to create 2D plans for plotting. Optionally, the model is analyzed, animations are generated, and the 3D model is rendered.

Step 1: **Start With a Template**

When Inventor starts, the New File dialog box appears.

- *To start new models:* select the *Standard.ipt* part template file, and then click **OK**.

Click **OK**, and Inventor opens in its Sketch mode of the Part environment, illustrated below.

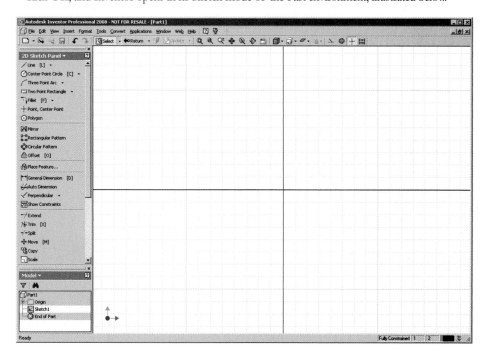

Step 2: **Draw 2D Sketches**

Even the most complex of Inventor's 3D models begin a simple sketch. These are made of circles, lines, and other 2D objects with which you are already familiar. The relationships between objects are determined by *constraints* (like object snaps, but with glue). The sizes of the sketch objects are specified later by dimensions.

- *To start new 2D sketches:* after you open a new part template, Inventor automatically enters 2D sketch mode.

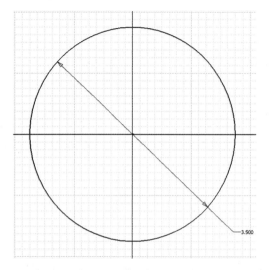

2D sketches are built up with drawing and editing commands, such as Circle, Fillet, and Offset. Inventor's sketching and editing commands are shown in the *panel* ("dashboard") illustrated at right.

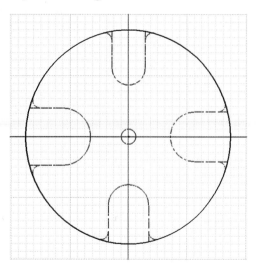

The sketch forms the platform for the next stage in 3D design, the *part*.

Step 3: **Extrude to Create Parts**

Inventor converts 2D sketches to 3D parts through *features*, such as extrusions, revolves, and sweeps. It can apply 3D fillets, drill holes with threads, add bends, and so on. Inventor's part editing commands are shown in the panel at right.

- *To create parts:* click **Return** on the Standard toolbar. This action exits Sketch mode, and returns to Inventor's Part environment.

At this stage, the model is an accurate 3D representation that permits Inventor to determine mass properties and perform engineering calculations.

Part 4: **Collect Parts into Assemblies**

For complex models, you collect parts into *assemblies* using constraints, mates, and other relationships. Inventor's assembly commands are shown at right.

- *To collect parts into assemblies:* from the **File** menu, click **New**, and then choose the *standard.iam* assembly template file. In the **Assembly** panel, choose **Place Component**. After placing components (parts and other assemblies), use the **Constraint** command to attach them together.

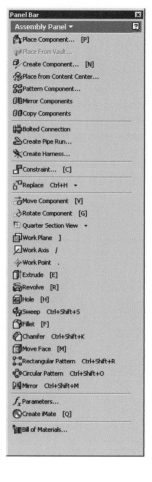

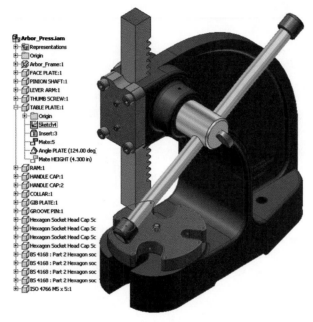

Left: *Model tree listing names of parts used by the assembly.*
Right: *Assembly model.*

At this point, the 3D model is complete, but further steps are possible. From this model, you can derive renderings, BOMs, stress analyses, and 2D production drawings.

Step 5: **2D Drawings from 3D Models**

Inventor easily creates 2D plans of 3D parts and assemblies in its Drawing environment, the equivalent to AutoCAD's layout tab. You position views (viewports), and then Inventor automatically generates the 2D and isometric views — in monochrome or color. You can also place parts lists, tables, title blocks, dimensions, balloons, and other annotations. Inventor's view and annotation commands are shown at the bottom of the page.

- *To create 2D drawings:* from the **File** menu, click **New**, and then choose the *standard.idw* drawing template file. In the **Drawing Views** panel, choose **Base View**.

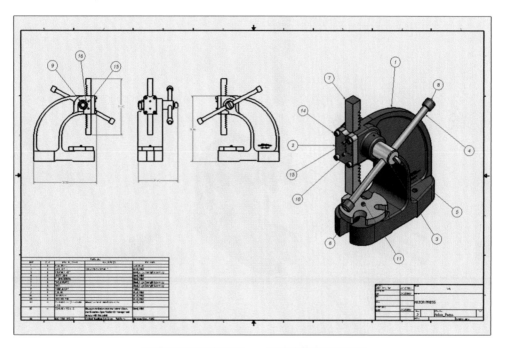

Step 6: **Generate Animated Presentations**

Inventor creates animations of the assembly's operation in its Presentation environment. It can also explode assemblies to show how the machines are to be assembled in the shop (illustrated below). Inventor's presentation commands are illustrated at the bottom of the page.

- *To create presentations:* from the **File** menu, click **New**, and then choose the *standard.ipn* presentation template file. In the **Presentation** panel, choose **Create View**.

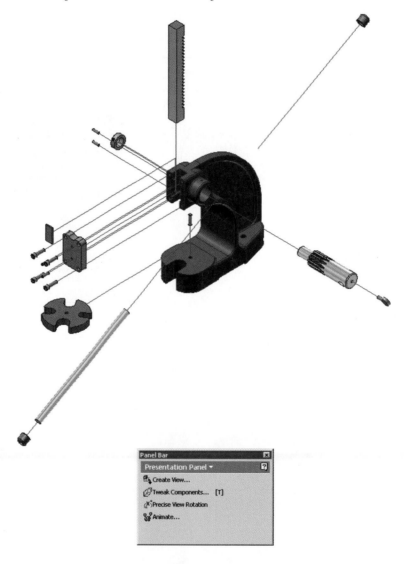

Step 7: **Render Models**

Inventor can create renderings in its Studio module. It produces photo-realistic renderings suitable for product brochures (left) and artistic renderings suitable for technical documentations (right). Rendering commands are shown below.

- *To create renderings:* in the Part or Assembly environment, choose **Inventor Studio** from the **Applications** menu. In the **Inventor Studio** panel, choose **Render Image**.

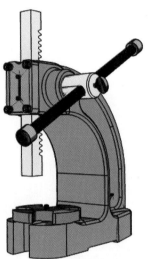

Design Assistants & Speciality Models

Inventor has assistants and modules for creating specialized 3D models and 2D drawings, as illustrated on the following pages. Design Accelerators are available in all versions of Inventor; Tube& Pipe and Cable & Harness are in Inventor Routed Systems Suite and Inventor Professional; Dynamic Simulation and FEA are in Inventor Simulation Suite and Inventor Professional.

Design Accelerators

Inventor provides Design Accelerators for performing engineering calculations and creating standards-based geometry. After you enter the design requirements, the generators and calculators create mechanically correct components automatically. See the boxed text on the following page for the complete list of accelerators.

- *To access design accelerators:* open an *.iam* assembly file, and then choose **Design Accelerator** from the **Assembly Panel** droplist.

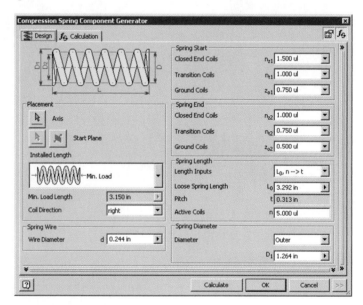

Left: Design Accelerator panel.
Right: *Design Accelerator for designing coil springs.*

Spring and gears designed by Design Accelerator.

DESIGN ACCELERATOR SUMMARY

Inventor's Design Accelerators consist of component generators, mechanical calculators, and *The Engineer's Handbook*.

Component Generators

The component generators create the following mechanical devices:

- Bolted connections.
- Shafts.
- Involute and parallel splines.
- Key connections.
- Disc and linear cams.
- Spur, bevel, and worm gears.
- Bearings.
- V- and synchronous belts.
- Roller chains.
- Clevis, joint, secure, cross, and radial pins.

Engineering Calculators

The engineering calculators solve the following problems:

- Plain bearings.
- Plug, groove, butt, spot, filled, and fillet welds.
- Butt, bevel, lap, step tube, and step solder joints.
- Separated hub, slotted hub, and cone joints.
- Tolerances.
- Limits and fits.
- Press fits.
- Power screws.
- Beams and columns.
- Plates.
- Shoe drum, disc, cone, and band drum brakes.

The Engineer's Handbook

The Engineer's Handbook is a help file containing formulas useful to mechanical engineers, such as the one illustrated below.

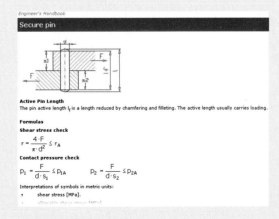

<tabular>Engineer's Handbook</tabular>

Secure pin

Active Pin Length
The pin active length l_f is a length reduced by chamfering and filleting. The active length usually carries loading.

Formulas
Shear stress check

$$\tau = \frac{4 \cdot F}{\pi \cdot d^2} \leq \tau_A$$

Contact pressure check

$$p_1 = \frac{F}{d \cdot s_1} \leq p_{1A} \qquad p_2 = \frac{F}{d \cdot s_2} \leq p_{2A}$$

Interpretations of symbols in metric units:
τ shear stress [MPa].

Dynamic Simulation

Inventor can simulate the dynamic motion of assemblies after you (1) define mechanical joints between parts, and (2) add forces on the parts. To simulate motion accurately, Inventor uses the mass, inertia matrix, and mass center position associated with parts. The resulting motion can be shown in a graph and exported as a movie.

- *To access dynamic simulation:* open an *.iam* assembly file, and then choose **Dynamic Simulation** from the **Applications** menu.

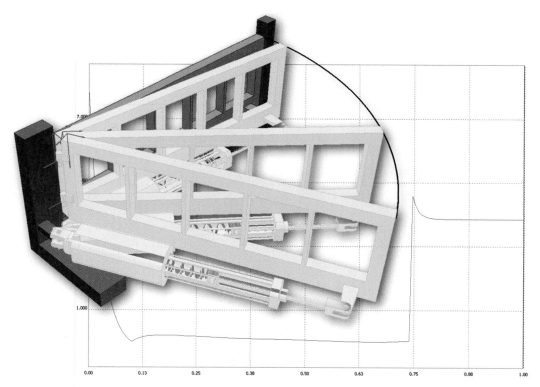

Left: *A representation of dynamic simulation...*
Right: *...superimposed on an output graph.*

Stress Analysis

Inventor can calculate the stresses that parts will probably undergo. To do so, you apply loads and constraints, and then have Inventor determine how the part will behave. Reports and movies can be generated from the analysis. (Currently, Inventor uses finite element analysis software from ANSYS; with Autodesk's purchase of PlassoTech, I anticipate the FEA program will change with the next release of Inventor.)

- *To access stress analysis:* open an *.ipt* part or *.iam* assembly file, and then choose **Stress Analysis** from the **Applications** menu.

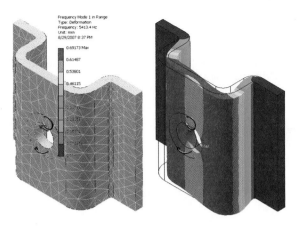

Left: *A part defined by meshes.*
Right: *The result of the stress analysis shown by colors representing the amount of strain.*

Sheetmetal

Inventor can optimize the layout of sheetmetal parts, which converts folded model (Folded Model environment) to flat patterns (Flat Pattern environment) for manufacturing. Sheetmetal parts have consistent parameters, such as constant thickness of material, relief size, and bend radius.

- *To access sheetmetal design:* select the Sheet Metal template in the New File dialog box. Upon returning from the Sketch environment, the Sheet Metal panel appears.

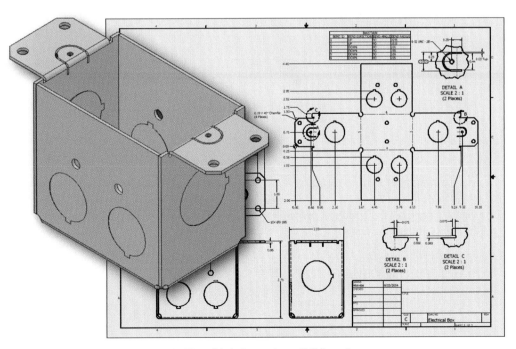

Left: *3D model of sheetmetal part (folded metal).*
Right: *Drawing of flattened sheetmetal (flat pattern).*

Wire Harness

Inventor can design harness assemblies with wires, cables, and connectors, which are placed in assemblies. It automatically calculate lengths and bundle diameters. To help with the design, Inventor can import electrical connectivity wire lists from *.xml* and *.csv* files, and then export a description of the completed design to *.xml* files. Connectivity lists are generated from ladder logic or schematic diagrams from the related AutoCAD Electrical software.

- *To access wire harnesses:* open an *.iam* assembly file, and then choose **Create Harness** from the **Assembly** panel.

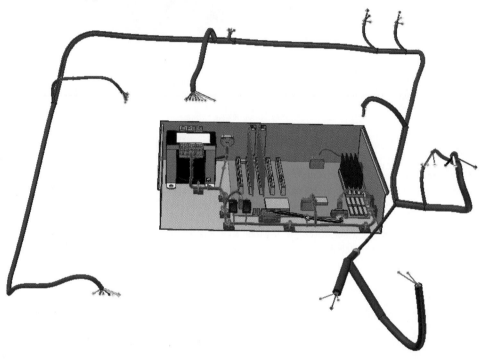

Outside: Wire harness with cables.
Inside: Computer case containing the harness.

Nailboard

The Nailboard environment creates 2D flattened views of the wiring harness as drawings for their manufacture. The drawings show wires as straight lines and ribbon cables as rectangles, and include bills of materials, wire lists, images of connectors, dimensions, and pin numbers.

- *To access nailboard drawings:* open an *.idw* drawing file, and then choose **Nailboard View** from the **Drawing Views** panel.

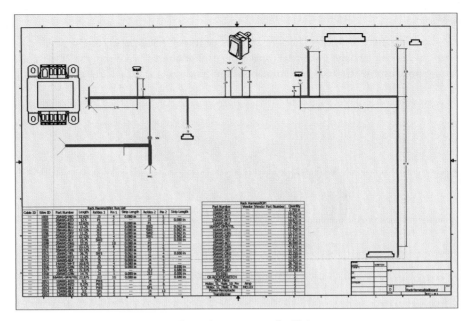

Nailboard view of a computer case's cable harness.

Frame Generator

Inventor uses the Frame Generator to draw frame assemblies for machines. After drawing the skeleton as a part, place it in an assembly file, switch to Frame Generator mode, and then insert frames.

- *To access frame generator drawings:* open an *.iam* assembly file, and then choose **Frame Generator** from the **Assembly Panel** droplist.

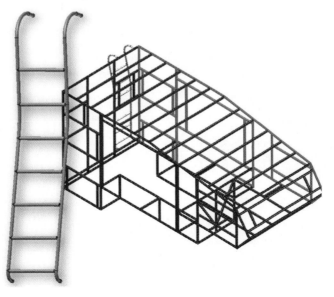

Left: *Ladder made as a frame.*
Right: *Camper structure made from several frames.*

Weldment

Inventor can specify weldments either as solids or as cosmetic features. Solid welds can be fillet or groove welds; cosmetic welds look nice and take up less space, but cannot be analyzed. From the 3D assembly, Inventor extracts welding symbols and bead annotations for drawings.

- *To access weldments:* convert an assembly to a weldment assembly using the **Convert** menu.

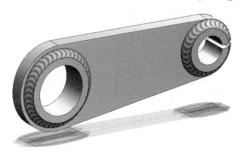

Welds attaching the cylinders to the link.

Tube and Pipe

Inventor can design tube and pipe runs of rigid pipes, bent tubes, and flexible hoses. These can have branches, connectors, and valves. The 3D Orthogonal Route tool creates intermediate route points at specific positions and distances. Inventor determines route type, nominal diameter, minimum and maximum segment length, material, bend radius, and other details.

- *To access tube and pipe models:* open an *.iam* assembly file, and then choose **Create Pipe Run** from the Assembly Panel droplist.

Steel and rubber pipes with bends, connectors, and valves.

The Benefit of History-based Design

In the MCAD world, there is much discussion over the pros and cons between history-based design and its competitor, freeform design. The bestselling CAD packages, including Inventor, use history, which in itself suggests that users prefer structured design over unstructured. The slower-selling CAD packages that advocate unstructured design maintain that their software makes it easier to make unanticipated design changes. The problem they have, however, is that the designer's intent is unclear, because there the history is missing.

Let's take an example. I've inherited the drawing of a clevis, without any information on how the design was created. If, however, the drawing file provided to me is in Inventor format, I can replay the work of its designer through the history tree — called "Model browser" by Inventor.

Examining Sketch 1

To see all of the history tree, I expand it by right-clicking the file name ("assign-03.ipt"), and then selecting **Expand All Children** from the shortcut menu. Models begin with sketches, and the first sketch is named "Sketch1." (If it has been renamed, then it is the first sketch listed by the tree.)

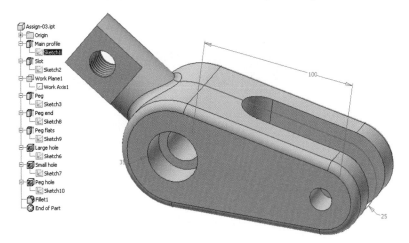

Inventor highlights the sketch and shows me the dimensions. From how the sketch is centered in the clevis, it appears to me that the designer did a double-sided extrusion of the sketch, and then later applied holes and fillets.

TIPS *If I wanted to edit the sketch at this point, I would right-click **Sketch1**, and then select **Edit Sketch** from the shortcut menu. This opens the sketch in Inventor's Sketch environment (see figure below). Any changes I make here are immediately reflected in the 3D model.*

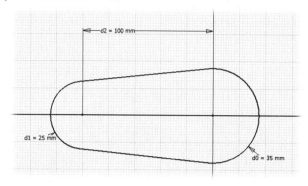

*If the 2D sketch appears in a strange 3D viewpoint, I use the **View | Look At** command to have Inventor change to plan view.*

*To return to the 3D model, I click the **Return** button on the toolbar, or else press **Ctrl+Enter.***

To check my hunch, I move the cursor up to the parent of Sketch1, "Main Profile" (I don't click Main Profile, but just rest the cursor over it so that I see both the sketch and the related profile.) I see that the width of the clevis was generated from Sketch1. My hunch is correct.

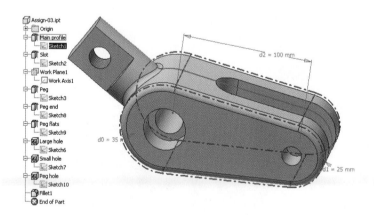

Examining Sketch2

Let's see what the next sketch tells me of the designer's intent.

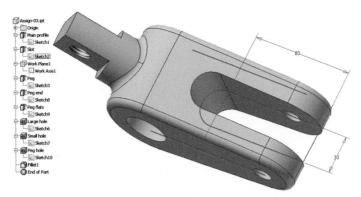

Sketch2 defines the cutout of the clevis. I move my cursor up to **Slot** to see the 3D slot generated by the 2D sketch.

The Work Plane & The Peg

The next item in the history tree is named "Work Plane." This tells me the designer defined a work plane before starting work on the clevis' peg. (*Work planes* are likes UCSs in AutoCAD.) To determine its angle, I right-click **Work Plane** and then choose **Show Dimension** from the shortcut menu. The angle, -60 degrees, appears in a small dialog box. I click the check mark to close the dialog box.

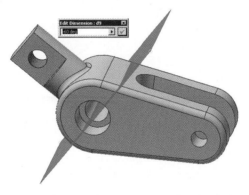

By clicking **Sketch3** and then hovering the cursor over Peg, I see that the designer drew a circle on the angled work plane (shown in blue), and then extruded it to create the peg. The extrusion limits are shown in red.

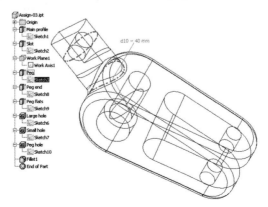

TIP *To make it easier to see inside the model, I change its visibility to* **Wireframe** *display.*

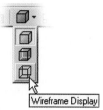

Wireframe Display

Examining the Fillets

I continue working my way through the history tree, until I arrive at the end. The designer's final step was to add fillets to the clevis. If I want to change the radius of the fillets, I can right-click **Fillet1** and then select **Edit Feature** from the shortcut menu.

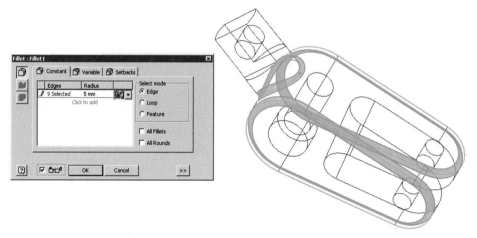

Inventor highlights the fillets, and opens a dialog box that gives me their details. When done, I click **OK** to dismiss the dialog box, and then click **Return** to return to the model.

By following the history tree, I was able to discover the designer's intent, and make modifications to suit my new requirements. This process is simply not possible with freeform 3D CAD packages, which lack the invaluable structure of the history tree.

(In Chapter 4, you have the opportunity to create a 3D model of the clevis discussed here.)

Before Installing Inventor: Computer Requirements

Installing Inventor is straightforward, but if you are not sure whether your computer has the needed horsepower to run Inventor, then visit the Autodesk Inventor Hardware Web Site at www.inventor-certified.com/graphics/index.php. It has a massive amount of information on the hardware and software needed to run Inventor 10, 11, and 2008. This site also provides help for problems you may encounter during installation.

You may be wondering if you need an especially hot computer to run a program as "big" as Inventor. Perhaps not. I'm running Inventor 2008 on a six-year-old computer that has these specs:

- **CPU**: 2.4Ghz Pentium 4.

- **RAM**: 1GB.

- **Free Disk Space**: About 500GB disk space divided between two internal and two external drives connected through FireWire.

- **Graphics Board**: nVidia FX 5500 connected to a recently-purchased 21" widescreen LCD monitor.

- **Operating System**: Windows 2000 updated to SP4, service pack #4.

My computer may be old, but it still works with Inventor for two reasons. When I handbuilt this computer back in 2001, I picked out components that were top-end for the time. And so this computer continues to work well with today's software, because hardware really hasn't changed much in the last half-decade.

Consider these points:

- CPU speeds have plateaued at just over 3GHz.

- Most software cannot take advantage of the dual-core and 64-bit CPUs that are common today.

- The limits inherent in 32-bit CPUs, motherboards, and operating systems cause RAM capacity to max out at 2GB, or 3GB under special conditions.

- Hard drives are bigger and cheaper, but no faster.

- The latest operating system from Microsoft, Vista, is too new to know if it speeds up or slows down your computer. (At time of writing, Autodesk felt that Inventor will be slightly faster on Vista because the Direct3D implementation is faster. Whether you'll notice this is hard to predict, due to the many other factors in Vista, such as the Aero interface and intensive digital-rights management.)

> **TIP** The files in the *<OS Drive>:\Documents and Settings\All Users\ Application Data\Autodesk\Software Licenses* folder must never be altered or deleted. Any change will trigger failure of your Autodesk licenses. You will then be required to reauthorize all impacted products. Do not restore your hard drive without first exporting the licenses with the Portable License Utility.

Hardware Recommendations by Autodesk

When it comes to Inventor, the most important piece of hardware is the graphics board and its device driver, because Inventor makes much use of the board's innate abilities.

Suitable Graphics Boards & Drivers

Just about any graphics board sold today meets minimum requirements to work with Inventor. The primary exception are graphics boards found on very low cost notebook computers, which tend to have insufficient RAM to run any graphics-intensive program.

Autodesk lists 239 models of graphics boards that it has tested to four levels of quality of with OpenGL and DirectX and multiple versions of Windows:

Autodesk — Autodesk and the graphics board vendor recommend this graphics board.

Green — all of Inventor's requirements are supported by the graphics board.

Yellow — some of Inventor's graphics requirements are not provided by the board.

Red — hardware acceleration does not work with Inventor; Inventor will run in slower software emulation mode.

The following are brand and product names tested by Autodesk for Inventor 2008:

ATI
> FireGL (*this is the only board ATI supports for CAD use*)
> All-in-Wonder
> Radeon
> Mobility FireGL

Intel
> 915G

Matrox
> Millennium

NVIDIA
> Quadro (*this is the only board nVidia supports for CAD use*)
> GeForce

S3
> Graphics Chrome

Sometimes similar sounding model numbers work quite differently. For example, nVidia's Quadro4 750 XGL rates a green checkmark, but the similarly named 780 XGL rates a red X. (In my computer, nVidia's GeForce FX 5500 is top-rated by both Autodesk and nVidia on Windows 2000 and XP. And to think I got it free from my son.) Thus, I urge you to visit www.autodesk.com/us/inventor/graphic_cards and check the qualifications of the graphics board in your computer.

Drivers

To make Inventor really shine, the device driver needs to be optimized, and so Autodesk lists version numbers of drivers optimized for Inventor for each operating system and graphics pipeline system (OpenGL or DirectX).

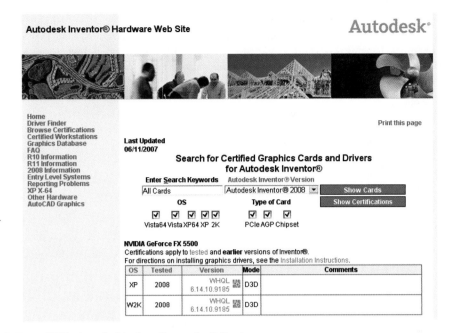

For my GeForce 5500, Autodesk's site tells me the following:

- Works with Windows 2000 and XP, but not Vista.
- Works with all aspects of Inventor 2008.
- Recommended by Autodesk and nVidia.
- Must run in DirectX mode (D3D).

My Windows 2000 system should be running drivers with version number 6.14.10.9185. To determine the version number in my computer, I take these steps:

1. Right-click the desktop, and then choose **Properties** from the shortcut menu.
2. Choose the **Settings** tab, and then click the **Advanced** button.
3. Choose the **Adapter** tab, and then click the **Properties** button.
4. Choose the **Driver** tab, and then examine the Driver Version number.

(MHQL is short for Microsoft Windows Hardware Quality Labs.)

If the version number does not match that recommended by Autodesk, decide whether you want to install it or not. Quite frankly, if Windows and Inventor and other software seem to be running fine, then don't change the driver. Sometimes changing drivers just causes more grief. On the other hand, if Inventor is not generating all the visual effects it can, then by all means download the appropriate driver from Autodesk's Web site.

Inventor 2008 uses either Microsoft's proprietary Direct3D or the open OpenGL graphics system to generate its 3D graphics quickly.

- **DirectX 9** is the default, and the only one available on Vista.

- **OpenGL** is available for Windows 2000 SP4 and XP SP2.

> **TIP** Inventor 2009 will support DirectX version 10.

Suitable Computer Components

If you plan to purchase a new computer for Inventor, look for one with these specifications in mind:

- *Fastest* 64-bit CPU.

  The CPU should have the *largest* L2 (on-board instruction) cache you can afford.

- Minimum 512MB RAM, but at least 1GB is recommended. Inventor works best with 3GB, which is available with XP SP2 and Vista. As soon as Inventor is available in a native 64 bit version, running 64-bit versions of XP or Vista will allow Inventor to access almost 4GB of memory — a gain of 33%. This can make a difference when working with large assemblies.

  If you need to trade off between the price of a fast CPU or memory (RAM), then Autodesk recommends that you purchase more memory. More RAM helps Inventor more than does a faster CPU.

- Minimum 80GB hard drive.

  Following installation, Inventor needs at least 1GB of disk space for its undo and temporary files. These files are kept in a folder pointed to by the operating system's TEMP environment variable. You may need to erase leftover files from time to time; only do this when Inventor is not running.

  A further 4GB free disk space is sometimes needed when the Content Center's SQL database needs to be repaired.

- Large virtual memory swap file.

  The swap file should be twice the size of physical memory. For example, a computer with 1GB RAM should have a 2GB swap file. If your computer is using the 3GB option, however, set the swap file to 4GB.

  You change the size of the swap file through Control Panel > System > Advanced > Performance > Settings > Advanced > Change in XP. In Windows 2000, change it through Control Panel > System > Advanced > Performance Options > Change.

- Pointing device, like a mouse or 3D controller.

  If a Logitech mouse operates slowly in Inventor, turn off the "Disable acceleration in games" option in the Mouse section of Control panel.

Autodesk has a list of low-cost computers suitable for educational use with Inventor at www.inventor-certified.com/graphics/value.php. They include models from Acer, eMachines, and HP.

> **TIP** *The following software may conflict with Inventor:*
> - *When Inventor's setup program asks for access to the Internet, grant it. Anti-spyware, anti-virus, and Internet security programs may attempt to prevent it accessing the Internet.*
> - *BlackIce firewall software may prevent the riched20.dll from being installed on your computer. This is an issue only if you install Mechanical Desktop using a deployment image, so turn off Blackice for the duration of the install.*
> - *Advertisement cannot install Inventor.*
> - *Internet Explorer prior to v6 may have problems with Inventor's Help system. Upgrade IE to v6 or later.*
> - *PDF Complete add-on for Office prevents Excel from opening when requested by Inventor.*

Installing Inventor

Follow these steps to install Inventor:

1. Before installing Inventor, direct **Windows Update** to install all critical updates from Microsoft. This increases the security of the operating system.

2. Check that **File and Printer Sharing** is turned on. This is needed for Content Center's ASDMS.

3. If your computer cannot install software from a DVD, request Inventor on CD format from Autodesk.

4. Check that your computer's graphics board and driver are supported by Inventor at www.autodesk.com/inventor-graphic-cards. If necessary, download and install the driver software recommended by Autodesk.

5. Turn of anti-virus software to help speed up the installation process.

6. Ensure that your computer has a connection to the Internet. During installation, allow the firewall software to grant access to Inventor's setup program. This will occur numerous times during installation.

7. iParts, iFeatures, thread customization, and spreadsheet-driven designs require Excel XP or later, which you may need to purchase and install.

8. Insert Inventor's DVD 1, and then run the *setup.exe* command. If there are any errors during setup, reboot the computer and try again.

9. Install everything from the DVD, because it all takes up less than 5GB of disk space — a pittance nowadays.

10. When installation is finished, start Inventor by clicking its icon on the desktop.

Autodesk Inventor
Professional 2008

TIP *Inventor turns off hardware acceleration by default. This is to ensure that Inventor works with any graphics board. To improve the speed of the graphics display, follow these steps after Inventor starts:*

1. *From Inventor's **Tools** menu, choose **Application Options**.*
2. *Select the **Hardware** tab.*
3. *Choose either the **Direct3D** or **OpenGL** option, depending on which provides better performance with your graphics board. If you are not sure, click the **Diagnostics** button for a report on the graphics board.*

   You can also choose according to your computer's operating system:
   - ***OpenGL** for Windows 2000 or XP.*
   - ***Direct3D** for Windows Vista.*

Getting Assistance

Whereas AutoCAD has a single menu listing online help, Inventor has two: Web and Help. The Web menu has these items:

Autodesk Inventor — opens the Web browser with Autodesk's page for Inventor.

Manufacturing Community — opens the browser with Autodesk's page for mechanical Weblogs (free).

Streamline — opens the browser with Autodesk's page for Streamline, its mechanical-oriented online project management service (not free).

Team Web — opens custom Web pages containing iDrop-compatible parts; requires installation of an ActiveX component from www.autodesk.com/prods/idrop/download/idrop.cab (free).

Supplier Content Center — accesses millions of pre-drawn parts from Part Solutions in these categories: bearings, tools and tooling, structural, power transmission, pneumatics, plumbing, hydraulics, hardware, fittings, fasteners, enclosures, electrical, and DIN (German industrial standard).

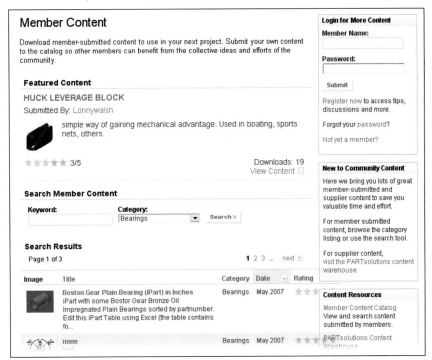

Help Menu

The Help menu lists the following items:

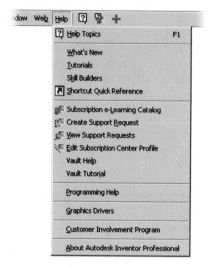

Help Topics — opens Inventor's online help window.

What's New — describes features new to this release of Inventor.

Tutorials — lists dozens of tutorials for using Inventor, such as building sheet metal parts and designing v-belt connections.

Skill Builders — accesses two dozen online tutorials located on Autodesk's Web site. These can be printed out, or read online.

Shortcut Quick Reference — displays Inventor's keyboard shortcuts, such as E for Extrude and F6 for isometric view. This list can be copied to the clipboard; a Customize key allows you to add and change shortcut definitions.

Programming Help — opens online help for working with Inventor's API (application programming interface).

Graphics Drivers — accesses Autodesk's Web page for checking graphics boards and drivers.

Customer Involvement Program — invites you to allow Autodesk automatically to collect information on how you use Inventor. The default setting is, "No, I don't want to participate at this time."

Menu Bar Buttons

The menu bar contains three buttons that also provide help:

Help Topics — opens Inventor's online help window; this is the same as pressing F1 or choosing Help Topics from the Help menu.

Visual Syllabus — lists animated tutorials.

Recover — runs the Design Doctor wizard, which attempts to solve errors in designs.

Online Discussion Groups

One of the better places to get quick assistance for free is through the discussions groups hosted by Autodesk. The groups for Inventor and related products are located at discussion.autodesk.com/adskcsp/index2.jspa?categoryID=21&discommunity=mfg. At time of writing this book, some of the topics of discussion included:

- Can an already drawn sheet be inserted into existing sheets?
- Can we hide .DWG filetypes?
- Anyone able to help with properties link?

Hot Issues

If you come across a bug in Inventor, the first thing to do is to check support.autodesk.com. From the Knowledge Base droplist, select **Autodesk Inventor**. The resulting Web page lists Hot Issues, which are the bugs that most users are wondering about. For more specific help, enter a phrase in the Search Support Knowledge Base text entry box.

Recent hot issues include these:

- End of Evaluation Period error during DWGIN.
- Known issues with Inventor 2008 and Windows Vista.
- Cannot create or edit text in IDW files after Microsoft Hotfix KB918118 installation.

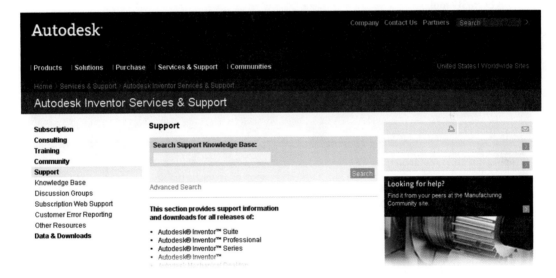

Web Blogs

Autodesk runs an umbrella blog for Inventor users called "Manufacturing Community" at mfgcommunity.autodesk.com. From here, you can access blogs written by Autodesk employees, such as these:

- Brian Roepke's *Under The Hood* on Vault, ProductStream, and more.
- Garin Gardiner's *In the Machine* on Inventor news and information.
- Nate Holt's *Controlling the Machine* on electrical design.
- Andrew de Leon's *Drawing the Machine* on mechanical drafting and design.

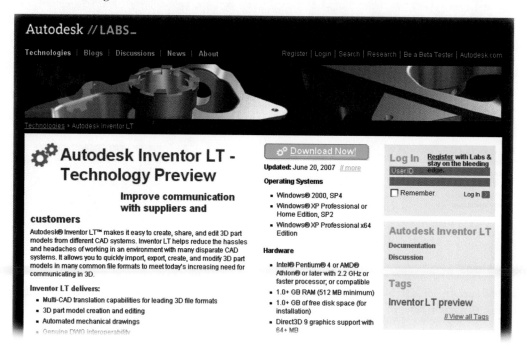

Not all beta software is kept under wraps through signed nondisclosure agreements. Autodesk Labs presents pre-release software for public consumption, including software add-ons of interest to Inventor users. At the time of writing this book, the labs.autodesk.com Web site offered these utilities for Inventor:

- Translator add-on.

- 2D-to-3D tool.

- Feature recognition tool.

Independent blogs include:

- Sean Dotson at www.sdotson.com.

- mCADForums at www.mcadforums.com.

- CAD Forum tips and tricks for Inventor at www.cadforum.cz/cadforum_en/tips.asp?CAD=Inv&CAT=CAD.

- Inventor PH at inventorph.wordpress.com.

User Groups

The Autodesk User Group International is the official user group for users of Autodesk software, including Inventor. Membership is free. Their Web site is at www.augi.com and their blog is at augi.typepad.com.

A Concise History of 3D at Autodesk

Over the last two decades, development of three-dimensional design has taken a tortuous path at Autodesk.

Beginning in 1982, Autodesk sold AutoCAD as an amazingly customizable CAD program, but limited to two-dimensional design. It was rumored that a full 3D system had been implemented in its core code, but had been abandoned when a key programmer left Autodesk to form his own company.

The first tentative step towards 3D was taken in 1985 when Autodesk added thickness and elevation, sometimes called "2-1/2D." Entities could be made 3D by giving them thickness and drawing them with a single z-coordinate (elevation). For instance, you could now draw a box by giving thickness to the lines making up a rectangle. At the time, we joked that it was easy to tell which skyscrapers had been designed in AutoCAD: they were the ones with flat tops.

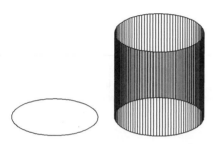 

Left: *Circle given thickness becoming a cylinder.*
Right: *Sphere made of meshes.*

To make a cylinder, you extruded a circle — but the cylinder had no top and no bottom. And you couldn't make spheres, not until Autodesk shipped AutoCAD Release 10 on 10/10/90. It was marketed as "The 3D Release." Nearly all entities could now be drawn with a z coordinate. No more flat-top skyscrapers. But the 3D was still simple, consisting of wireframe and mesh drawings only. For a while, there were workarounds, such as the 3dLine command for drawing lines in 3D. While this is excruciatingly primitive by today's standards, it was exciting at the time for devoted followers of AutoCAD, whose mantra was, "You're limited only by your imagination."

The first taste of solid modeling came with Release 10's tenuous link to AutoSolid, a $5,000-package that required the Xenix operating system. (The section following has more details on this detour in 3D software.)

Release 13 contained the most significant changes to AutoCAD ever: the aging CAD system became modern and object-based, its ObjectARX API allowed custom objects, the built-in rendering was free, and the kernel switched from the aging PADL to the now-standard ACIS.

Most significantly, solids modeling was no longer an extra-cost option, making it available to all AutoCAD users. There was much initial excitement over AutoCAD's free solids modeling. One prediction was that architects would use solids modeling for their massing studies. Another was that this heralded the era of Everything 3D, all the time.

Then with AutoCAD 2007, Autodesk added 3D surfacing, a simple version of history to solid models, and the ability to exchange drawings with Inventor; with 2008, AutoCAD was able to exchange Inventor's DWG drawings without translation.

But even with Release 13's spectacular improvements, the shortcomings of AutoCAD's limited 3D capabilities quickly became apparent to hardcore mechanical engineers, and Autodesk knew it needed to try a different approach.

Before Inventor: Autodesk's First Solid Modelers

While AutoCAD users of 1990 were excited by AutoCAD Release 10's 3D abilities and its increasingly powerful AutoLISP programming language for manipulating 3D entities, Autodesk was looking toward the future of 3D: solids modeling. This technology could simulate the real world, and allowed new products to be designed, tested, and visualized with no need for physical prototypes. The potential in cost and time savings to users was enormous, as was the potential for hugely increased revenue for Autodesk.

Autodesk took a first tentative step in early 1987, spending $6 million on Cadetron of Atlanta, Georgia and its The Engineer Works 3D solids modeling software — tentative, because Autodesk eventually abandoned the software.

For its time, The Engineer Works was extremely powerful software, but cumbersome. Powerful, because it could do FEA (finite element analysis) on 3D solid models; cumbersome, because it ran only on the SCO"s Xenix 386 operating system. (Trivia: Xenix is a version of Unix developed by Microsoft for personal computers; it was the most popular variant of Unix in the 1980s. In 1987, Microsoft sold it to Santa Cruz Operation.)

The Engineer Works was based on the PADL-2 (Part and Assembly Description Language) modeling kernel developed a decade earlier at the University of Rochester, New York.

During the beta program, Autodesk renamed its solid modeling software "The Pioneer Program" and required that beta testers pay for it: $2,500, and they needed to pay $1,000 for the Xenix operating system, as well as several thousand dollars more for a 386 computer. This was a foreshadowing that solids modeling would not be cheap.

When Autodesk shipped the solid modeler in 1988, they named it "AutoSolid" and priced it at $5,000. A screen grab of the program can be viewed at members.ozemail.com.au/~cadwest1/gallery/parker.jpg. But it did not sell well, and so a year later Autodesk reworked the modeler to run on DOS as an extension to AutoCAD Release 11. Again it was renamed, this time as Advanced Modeling Extension (AME), and the price was cut to just $500.

One of Autodesk's founders, John Walker, wrote at the time, "We tried one thing with AutoSolid, and it didn't work. Now we're going to try something else [AME], preserving most of work we've poured into it so far, and I think we'll end up taking over the solid modeling business, in time" (ref: www.fourmilab.ch/autofile/www/chapter2_77.html). He made this predication in 1989.

A decade later, Autodesk launched Inventor, and then six years after that, the prediction came true when Autodesk took the lead in the solid modeling market, in terms of unit sales.

During the intervening decade, however, the company took more detours to arrive at its position of leadership.

Autodesk undertook a number of additional acquisitions to solve the problem of selling mechanical design software that worked well. Woodbourne Design Companion was a remarkable 2D add-on to AutoCAD that quickly disappeared, but this acquisition had an everlasting impact on Autodesk. The head of Woodbourne Software was Robert "Buzz" Kross, who became the vice president of Autodesk's mechanical division and implemented Walker's dream of dominance in the solid modeling market.

1992 — Autodesk spent $15 million on Micro Engineering Solutions Inc. of Novi Michigan, and its Solution 3000 surfacing and machining software. Autodesk still maintains a development office there in automotive country. A year later, Solution 3000 was rebranded AutoSurf and became an add-on for AutoCAD, giving it freeform NURBS (Non-Uniform Rational B-Spline) curves and modeling. That ended development on AME.

1993 — Autodesk acquired Woodbourne of Lake Oswego, Oregon and its Design Companion software. This software eventually became Mechanical Desktop.

1994 — Autodesk released AutoCAD Designer, its first feature-based parametric solid modeler. It was an add-on to AutoCAD Release 12 and priced at $1,500.

In addition to Designer, the mechanical software lineup now included the following:

AutoSurf, an AutoCAD extension for 3D surfacing and machining.

Aemulus, an AutoCAD extension for reading CADAM files, a mainframe-based solid modeler developed by Lockheed.

ManufacturingExpert, a NURBS (non-uniform rational B-spline) surface modeler integrated with 2-5-axis NC (numerical control) tool path generator for CNC (computer numerical control) mills.

At this early point, market research firm Dataquest already ranked Autodesk as the second largest MCAD (mechanical CAD) software company worldwide. Autodesk's CEO, Carol Bartz, was enthusiastic: "With nearly one million users already familiar with AutoCAD, we believe the move to working with 3D design tools within AutoCAD will be a natural transition. AutoCAD Designer and AutoSurf Release 2 will transform the MCAD marketplace from 2D drafting and detailing to 3D design."

1996 — Autodesk released Mechanical Desktop 1.0 as an add-on to AutoCAD, priced at $4995 (including AutoCAD). Autodesk felt that MDT would continue to hold this advantage over competitors: a user interface already familiar to a million AutoCAD users. But being tied to AutoCAD was a serious disadvantage, because it was never designed for history-based parametric solids modeling. At some point in the next two years, the decision was made to write a whole new MCAD system from the ground up.

1998 — Autodesk purchased software from Genius CAD-Software for $69 million, including Genius AutoCAD, Genius AutoCAD LT, Genius Desktop, the Genius Vario parametric modeler, and Genius Modules. Some of this software was integrated into Mechanical Desktop, while the remainder was repackaged into Power Packs.

The same year, Autodesk launched AutoCAD Mechanical as a 2D-only mechanical design add-on to AutoCAD Release 14. Version 1 was numbered "Release 14.5," because that matched the version number of Genius Mechanical software upon which it was based.

1999 — Following months of industry rumors over a mystery software project code-named "Rubicon," Autodesk launched Inventor as its AutoCAD-independent 3D parametric modeling software. Rubicon is the river in northern Italy that Julius Caesar crossed in 49 BC as a deliberate act of war against Gaul, modern day France and the head office location of Inventor's primary competitor, Dassault Systemes and SolidWorks.

2001 — After using the ACIS solid modeling kernel for several years in AutoCAD and Inventor, Autodesk exercised its right to buy version 7 from Spatial Technologies, owned by Dassault Systemes. From the source code, Autodesk developed its own kernel, named ShapeManager. The company explained it could no longer wait for Spatial to implement features needed for Inventor. Autodesk now had a single solid modeler that it alone controlled for use by AutoCAD, Inventor, and other products.

2002 — Autodesk bundled Mechanical Desktop with Inventor 5, naming the package "Inventor Series." (AutoCAD and Mechanical are also included in the box.) Even through it had sold 200,000 licenses, Autodesk decided to make MDT unavailable as a separate product — although 2D Mechanical continues to be available to this day. The purpose of the bundling was to convince MDT users to upgrade to the non-AutoCAD-based Inventor.

Brief History of Inventor

Inventor is a relative latecomer to the world of solids modeling. It was launched in the fall of 1999, years and decades after many other 3D solid modelers: Unigraphics (started in 1973; since renamed NX), Cadam (1977; merged into Catia), Cadkey (1984; renamed KeyCreator), ME10 (1984; renamed CoCreate), Pro/Engineer (1986), SolidWorks (1994), and Solid Edge (1996). The technical editor reminisces that Computervision's CADDS 4X and a host of other minor products have come and gone as well; at one time he probably could name 20 mechanical CAD packages.

To catch up with capabilities found in competitors' software, Autodesk initially updated Inventor twice a year; now it is released once a year in April, at the same time as other software from Autodesk.

Version	Release Date	Code Name
Automobiles:		
Pre-release	...	Rubicon (Chrysler Jeep)
Inventor 1	September 1999	Mustang (Ford)
Inventor 2	March 2000	Thunderbird (Ford)
Inventor 3	August 2000	Camaro (GM)
Inventor 4	December 2000	Corvette (GM)
Inventor 5	September 2001	Durango (Chrysler)
Inventor 5.3*	January 2002	Prowler (Chrysler)
Inventor 6	October 2002	Viper (Chrysler)
Inventor 7	April 2003	Wrangler (Chrysler Jeep)
Inventor 8	October 2003	Cherokee (Chrysler Jeep)
Inventor 9	July 2004	Crossfire (Chrysler)
Inventor 10	April 2005	Freestyle (Ford)
Inventors:		
Inventor 11	April 2006	Faraday (magnetic fields)
Inventor 2008	April 2007	Goddard (rockets)
Inventor 2009	April 2008	Tesla (alternating current)

Note:
) Introduction of ShapeManager.

During 2003-2007, Autodesk made a flurry of acquisitions to beef up the capabilities of Inventor:

Compass — data management software.

Engineering Intent — engineer-to-order (ETO) software.

Linus Technologies — cable and wire harness modeling automation software.

MechSoft Technology — 50+ calculators and design wizards, a 1.5 million-part drag-and-drop content library, and mechanical engineering handbooks.

Solid Dynamics — analysis of kinematics and dynamics in mechanical assemblies.

truEInnovations — truEVault workgroup-based file management for Inventor; becomes Vault.

Navisworks — Navisworks 3D model viewer of many file formats.

Opticore — interactive and realistic 3D digital product visualizations and presentations.

PlassoTech — stress and thermal analysis, and simulation software.

VIA Development — generating wiring and other logical diagrams of machinery systems.

To take advantage of these acquisitions, Autodesk now ships Inventor in a variety of configurations at a variety of price points:

Inventor LT
Creates parts and drawings only; cannot create assemblies (currently available from Autodesk Labs).

Inventor Suite
Bundles AutoCAD, AutoCAD Mechanical, Mechanical Desktop, Inventor, Vault, Autodesk Management System, Design Review, TrueView, Inventor View, Design Accelerator, ANSYS Stress Analysis, and Content Center.

Inventor Routed Systems Suite
Adds Cable Harness for designing wire harnesses and wiring systems, and Tube & Pipe for the creation of piping, tubing, and flexible hose systems.

Inventor Simulation Suite
Adds finite element analysis and dynamic simulation modules.

Inventor Professional
Includes both Routed and Simulation modules.

ProductStream
An add-on to Vault, adds revision and release control.

COMPANION PRODUCT SUMMARY

Autodesk provides a host of products to accompany Inventor. See <u>usa.autodesk.com/adsk/servlet/</u> <u>index?siteID=123112&id=8411653</u> for the full list.

For Conceptual Design

AliasStudio — designs products ranging from 2D sketches to production models.

ImageStudio — creates photo-real images of products for marketing, catalogues, trade shows and the Web before they are manufactured.

PortfolioWall — offers a design review tool for team members and customers.

For 2D Mechanical Design

Electrical — creates and modifies electrical controls designs (built on AutoCAD).

Mechanical — offers 2D mechanical engineering design and drafting (built on AutoCAD).

For Data Management

Productstream — offers automated release and change management with tracking of BOMs and related design information.

For Collaboration

Streamline — manages collaborative project management using Web-based tools for data sharing.

DWG TrueView — views and plots DWG and DXF files, and publishes them as DWF files (free).

DWG TrueConvert — translates AutoCAD-based drawing files between Release 14, AutoCAD 2000, 2000i, 2002, 2004, 2005, and 2006 (free).

DWF Viewer — views and prints 2D and 3D drawings (free).

Design Review — reviews, marks up, and round-trips designs in DWF format (free).

DWF Writer for 3D — publishes 3D design data in DWF format from non-Autodesk applications (free).

Visualization

VIZ — models, renders, and presents 3D designs.

3ds Max — offers 3D modeling, animation, and rendering for film and television visual effects and animation, game development, and design visualization.

Chapter 2

Get Your Feet Wet with Inventor

Now that you've installed Inventor, let me take you on a guided tour of the user interface and some of its functions. You won't build anything just yet. Instead, you'll open one of the sample drawings included with Inventor, and then get a feel for how the software works.

Starting Inventor

1. Start Inventor by double-clicking its icon on the Windows desktop.

Autodesk Inventor
Professional 2008

IN THIS CHAPTER

- Starting Inventor and touring the user interface.
- Editing parts editing, and accessing sketches and features.
- Viewing the ContentCenter.
- Inputting precisely and using mouse buttons.
- Zooming and panning.
- Setting up drawing sheets and section views.
- Adding dimensions, generating parts lists, and placing text.
- Plotting drawings.

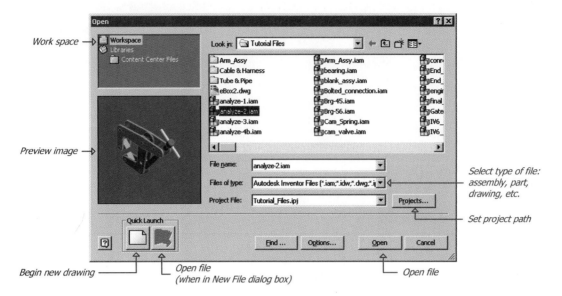

Work space →
Preview image →
Begin new drawing
Open file
(when in New File dialog box)

Select type of file:
assembly, part,
drawing, etc.

Set project path

Open file

Notice the Open dialog box. Above, I've annotated some of its buttons and windows.

2. Ensure that "Tutorial_Files.ipj" is selected for the **Project File**.

3. Choose *analyze-2.iam*. This is the sample drawing you'll work with in this chapter.

4. Click **Open**. Notice that the 3D model of the assembly appears in Inventor, as shown on the next page.

(Inventor can start in one of several drawing modes, called "environments." Because you opened an assembly, Inventor is now in its Assembly environment. The other primary ones are named Part, Drawing, and Presentation, and there are many others. You use some of these environments in this chapter, and learn about others later in this book. The content of Inventor's user interface changes according to the environment that's current.)

Touring the User Interface

Inventor looks vaguely like AutoCAD, and that's no accident. In AutoCAD 2007, Autodesk added the 3D workspace that was meant to mimic Inventor's user interface, especially with the sparse number of toolbars, visual styles, and the new Dashboard.

Let's take a look at how Inventor's user interface is similar to — and different from — AutoCAD. I'll start from the top, and then work my way to the bottom of the screen.

Menu Bar and Toolbars

The names on Inventor's menu bar share some similarities with AutoCAD, as the figures below illustrate:

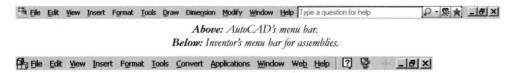

Above: *AutoCAD's menu bar.*
Below: *Inventor's menu bar for assemblies.*

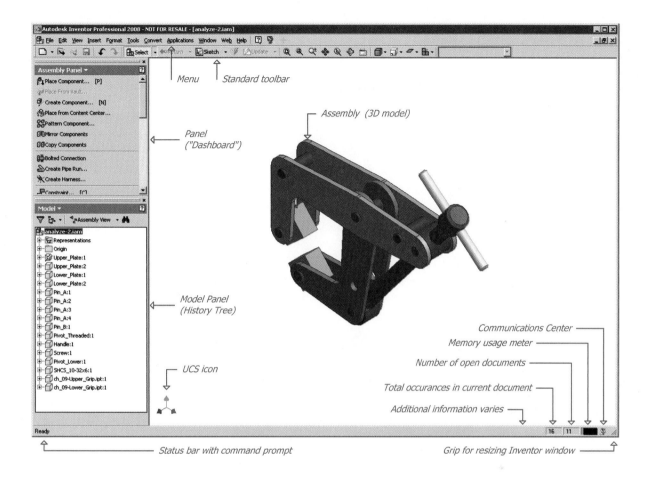

Menu   Standard toolbar

Assembly (3D model)

Panel
("Dashboard")

Model Panel
(History Tree)

Communications Center
Memory usage meter
Number of open documents
Total occurances in current document
Additional information varies

UCS icon

Ready

Status bar with command prompt   Grip for resizing Inventor window

The contents of the File, Edit, Window, and Help menus are very similar to those in AutoCAD; the other menus are very different, however. The items in menus change as you switch Inventor between environments.

Standard Toolbar

Like other Windows software, Inventor has a toolbar named "Standard." This toolbar provides access to standard functions, such as opening and saving files, and undoing mistakes. Many buttons on Inventor's Standard toolbar will already be familiar to you, but some are new to you.

The buttons on the Standard toolbar change according to the environment; the one illustrated above is for the Assembly environment. To access the other toolbars found in Inventor, right-click this one, and then select the toolbar names from the shortcut menu.

Sketch Properties Toolbar (Object Properties)

Inventor doesn't require you to control layers and other properties as you to in AutoCAD. Layers are assigned automatically to make it easier to translate 2D drawings with AutoCAD.

To change the color, linetype, and lineweight, use the Sketch Properties toolbar. It is the equivalent to AutoCAD's Object Properties toolbar. (To access it, right-click another toolbar and then select **Sketch Properties**.)

Bonus: unlike AutoCAD, Inventor always has all linetypes available; no need to load them!

The other toolbars are:

- **2D Sketch Panel** — equivalent to AutoCAD's Draw and Modify toolbars.
- **Inventor Precise Input** — equivalent to AutoCAD's command bar.

Third-party developers sometimes add toolbars and other controls to Inventor.

Panel (Dashboard)

The most prominent feature of Inventor's user interface is the Dashboard-like panel called the "Panel nar." Like AutoCAD's Dashboard, it consists of command-executing buttons logically grouped together. The Panel changes as environments change, just as AutoCAD's Dashboard changes according to the workspace. I show three of them below:

Left: *Panel with commands for manipulating assemblies.*
Center: *...for creating parts.*
Right: *...and for 2D sketching.*

I won't detail the Panel's functions here, because they are largely self-explanatory — other than to note that arrows indicate flyouts for command variations. In the figure below, the Center Point Circle button shows its flyout for accessing the Tangent Circle and Ellipse commands.

The flyout demonstrates something else: Inventor tends to have fewer variations of commands than does AutoCAD — as evidenced by its two methods for drawing circles, compared with AutoCAD's six.

> **TIPS** *Unlike AutoCAD, which tends to duplicate commands, Inventor segregates the functions in toolbars and menus: commands found in menus are not found in the Panel, and vice versa. The technical editor concurs: "To my knowledge there is virtually no overlap between the menus and the Panel."*
>
> *Many Inventor commands remain active, as in AutoCAD, until terminated with **Esc** or by another command.*

Browser Bar

A second panel displays the history of the parts being modeled. Autodesk calls this the "Browser Bar" — known as the "history tree" in other mechanical CAD packages. As you add features to the model, such as extrusions, fillets, and holes, the names are added to the tree in the order in which you created them.

(I think that Autodesk refers to the panel and the browser as "bars" because they act a lot like toolbars: they can float or be docked. Even if they don't look like bars.)

The tree is interactive with the drawing: when you select an item in the tree, it is highlighted on the model — and vice versa. Let's see how that works:

1. With the *analyze-2.iam* file open in Inventor, move the cursor over one of the gray L-shaped plates. Notice that it is outlined in red. This is how Inventor highlights parts.

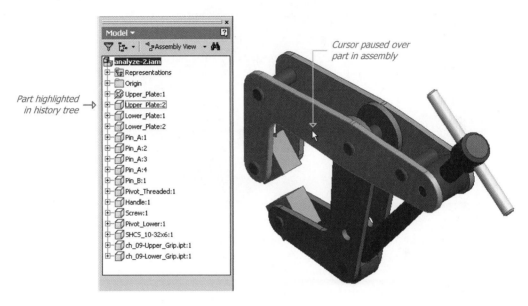

Part highlighted in history tree →

Cursor paused over part in assembly

2. Look over at the Model Browser. The name of the part is highlighted by a red box, "Upper Plate 2."

3. Move the cursor over one of the blue pins, and then pause for a second or more. Notice the icon that appears at the cursor. Its appearance means that the cursor is over two or more parts. Because the selection is now ambiguous, Inventor gives you the opportunity to choose the correct part.

- **Left and right arrows** — choose another part.
- **Green button** — choose the part highlighted in red.

4. Without moving the cursor, click either arrow until the L-shaped plate is again highlighted, and then click the green button. The plate is selected. (If you move the cursor at all, the selection mechanism resets itself.)

5. In the Model Browser, click the **+** next to the highlighted "Upper_Plate2". Notice that it opens to show you names like "Origin" and "Mate1." These are called "children." Origin is the local 0,0,0-coordinate for the part, while mates are connections.

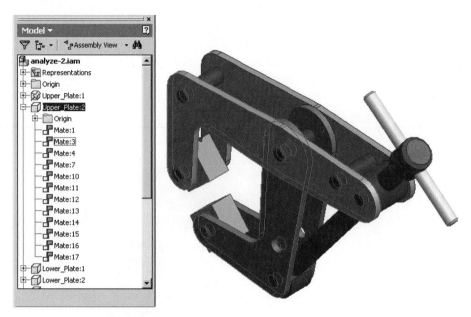

6. Move the cursor over the children, one by one. As you do, notice that the connections (*mates*) are highlighted. These show where Upper_Plate2 is connected to other parts in the assembly. (Some of the connections are point-to-point and so are not easy to see.)

Parts Editing

Let's carry on to see how parts are edited. Small correction: *part*, singular. You edit one part at a time in Inventor, although you can have many parts open for editing each in its own window.

(It is possible for Inventor to edit more than one part at the same time using adaptivity driven changes.)

There are two ways to edit a part in Inventor:

- **Edit** it in the Assembly environment. This lets you see the part in the context of the entire model. (To help reduce visual clutter, non-edited parts can be dimmed.)

- **Open** it in the Part environment. This opens the part in its own window, isolating it from the assembly.

Edit Parts

To edit a part in the Assembly environment, follow these steps:

1. Right-click the part, and then choose **Edit** from the shortcut menu.

Notice that the Panel and the Browser change their content. The Panel now lists 3D editing commands, like Extrude and Shell, while the Browser lists the features making up the plate, such as the extrusion and the holes. Data about parts not being edited are grayed out.

2. As before, you can run the cursor over the Upper_Plate:2's children to see the features highlighted in the assembly.

3. In the Browser, click the **+** sign next to Extrusion1 to expand it, and then click **Sketch1**. Notice that Inventor displays the dimensions defining the plate. Unlike AutoCAD, where dimensions are an afterthought, Inventor's dimensions truly define and control the sketches (2D outlines) and 3D features that define the part.

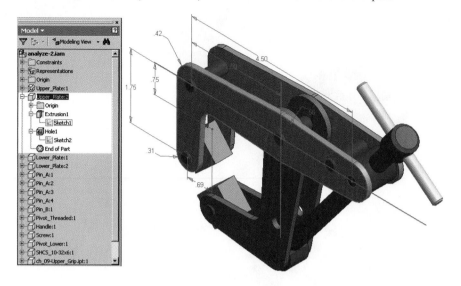

You'll leave parts editing for another chapter.

4. To exit this editing environment, click the **Return** button on the toolbar.

Notice that the Panel and Browser return to their assembly modes.

Open Parts (XOpen or BEdit or RefEdit)

To edit the part in the Parts environment, follow these steps:

1. Right-click the part, and then choose **Open** from the shortcut menu. Notice that Inventor opens a new window, isolating the plate for editing. This is similar to using AutoCAD's XOpen command to open a xref in its own window for editing, or BEdit to open a block in the Block Editor environment, or RefEdit to isolate a block for in-place editing.

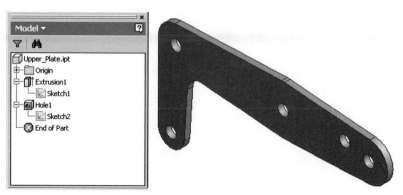

Again, notice that the content of the Panel and the Browser changes.

2. You are probably familiar with AutoCAD's grips for direct editing of objects. Inventor has them too, but Inventor's grips work in 2D and 3D. Let's see how these work:

Right-click the Extrusion feature, and then choose **3D Grips** from the shortcut menu.

Notice that the part takes on a 3D wireframe look.

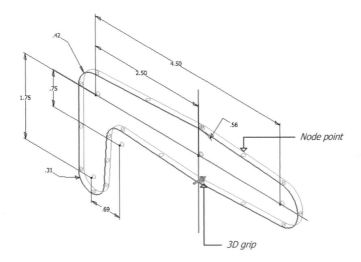

3. Dots appear; these are node points. Move the cursor over one.

As you do, notice that temporary 3D grips appear. The grip look like an arrow that points in the direction that editing can take place.

4. Click to place a 3D grip at a node point.

5. Drag the grip to resize the feature.

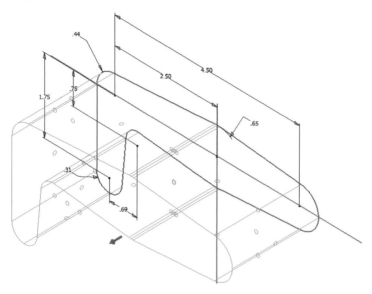

6. At any time, you can right-click the grip to access the shortcut menu with options for entering precise distances, such as offset and extents.

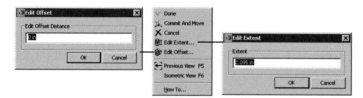

7. To end this portion of the tutorial, choose **Cancel** from the shortcut menu.

> **TIPS** *3D grips are able to change thicknesses, lengths, and radii of parts. Drag the grips to resize features and faces. For greater accuracy, grips can snap to other geometry.*

Accessing Sketches

Parts are made of sketches and features. You can burrow down one more level to the Sketch environment. With that step, you have moved Inventor from the *assembly* (the collection of 3D parts) to the *part* (a single 3D model) to the *sketch* (the 2D outline of the part).

Continuing with the tutorial drawing above, here are the steps:

1. In the Browser, open the "Extrusion1" tree, and then right-click **Sketch1**.

2. From the shortcut menu, choose **Edit Sketch**. (Alternatively, you can double-click **Sketch1** to directly open it for editing.)

Notice that the sketch opens in a new environment called "2D Sketch." You see the sketch for what it is: a 2D drawing that defines the shape of the part.

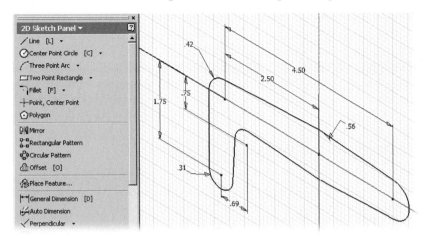

The Panel displays commands useful for 2D drafting, such as Line and Arc. In contrast, the Browser is no longer useful at this deep level of the model.

3. Probably, you are seeing the sketch displayed at an angle. To rotate it to plan view, use the Look At command, as follows:

   a. From the **View** menu, choose **Look At**.

   b. On the status bar, Inventor asks you:

   Select entity to look at

   c. Pick one of the lines in the sketch. Notice that it rotates smoothly into plane view.

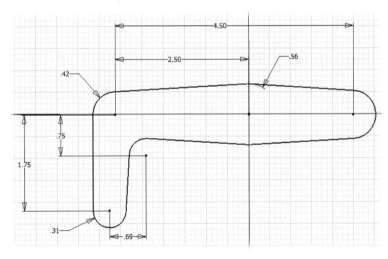

   TIPS *As an alternative, you can select the sketch first, and then click the **Look At** button on the Standard toolbar.*

   *If the **Parallel View on Sketch Creation** option is enabled in the Tools | Application Options | Sketch dialog box, the view automatically rotates perpendicular to the sketch plane.*

4. To exit the Sketch environment, click **Return** on the toolbar.

Accessing Features

There is one more environment I'd like to show you before returning to the tour of Inventor's user interface. It is the Feature environment. *Features* are 3D faces based on 2D sketches, extrusions being the most common type.

In the Features environment you perform most of the 3D editing tasks that you are familiar with from AutoCAD — revolve, extrude, 3D chamfer and fillet, and so on. Inventor lacks AutoCAD's commands specific to Boolean operations, such as Union and Subtract; instead, Boolean operations are part of feature creation commands, such as Extrude. In the Extrude dialog box below, the union (Join), subtract (Cut), and intersect (Intersect) operations are controlled by the three buttons stacked vertically. (Boolean operations are also supported via Derived Components.)

To get to the Feature environment, follow these steps:

1. In the Browser, right-click the name of a feature, such as "Extrusion."

2. From the shortcut menu, select **Edit Feature**. Notice that the Extrude dialog box appears. As well, the face of the feature is highlighted in cyan.

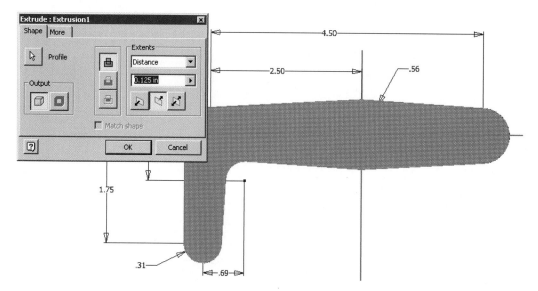

3. The part is still in plan view, which isn't a useful viewpoint for 3D editing. To return to the previous 3D viewpoint, follow these steps:

   a. Right-click a blank spot in the drawing window.

   b. From the shortcut menu, choose **Previous View**.

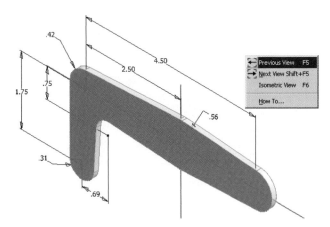

Notice that the plate smoothly rotates back to the previous 3D viewpoint. That's better: now you clearly see the base face in cyan and the extrusion (outlined in green).

4. The other features are holes, and you can edit them in a manner similar to the extrusion:

   a. Right-click one of the "Hole" entries in the Browser.

   b. Choose **Edit Feature** from the shortcut menu.

   This time a dialog box for defining holes appears, as do the dimensions that define distances between holes. This shows how Inventor tracks all the data needed to define parts and assemblies.

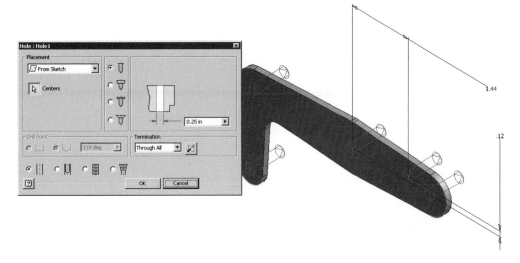

I should note that dialog boxes in Inventor don't act like dialog boxes in AutoCAD. Even though they have OK and Cancel buttons, use of the buttons is not mandatory: you can execute any other command while the dialog box is open. In that way, Inventor's dialog boxers act more like AutoCAD's palettes. You click OK and Cancel only when you no longer need the dialog box.

5. To exit this window used for editing parts (and sketches and features), close the window with the **File | Close** command. If you've made any changes, Inventor will ask if you want to save them.

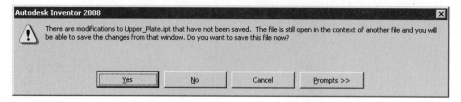

6. For this tutorial, click **No**. Notice that the Assembly environment returns.

With that in mind, let's return to our interrupted tour of Inventor's user interface.

Status Bar

You may find that the status bar is not as informative as you are used to from AutoCAD. For example, there are no buttons for turning the grid on and off quickly. (In Inventor, you have to change the grid through a dialog box. But in practice, the technical editor says he rarely ever uses the grid, not in Inventor nor in AutoCAD.) Indeed, the only button on Inventor's entire status bar is the one for the Communications Center.

The left half of the status bar displays Inventor's command prompts.

In the right half, the display of status bar varies, depending on what's happening in the model. For instance, while drawing a line you'll see the following data reported:

*   X,y coordinates (cursor position).
*   Length of line.
*   Angle of line.

Other data fields are always displayed, even when the drawing is empty:

*   Number of occurrences in active document.
*   Number of open documents or windows.
*   Memory consumption (green bar shows amount of real+virtual memory used).
*   Communications Center (connects with Autodesk through the Internet).

Precise Input (Command Bar)

When Inventor opens, one of the first things you may have noticed is that the command line is missing. That's because Inventor is more cursor-oriented than AutoCAD. (AutoCAD was written in the early 1980s when computers were largely keyboard-oriented — hence its emphasis on command-line input.) Nearly everything in Inventor is accessed by mouse, while command prompts are displayed in the status bar.

However, there may be times when you need to enter precise coordinates by keyboard. For these times, you open the Precise Input toolbar, as follows:

1. Right-click any toolbar.
2. Select **Inventor Precise Input** from the shortcut menu. Notice the Inventor Precise Input

toolbar. The content of this toolbar changes, depending on the drawing or editing command in effect. The figure below illustrates just one example:

However, the precise input bar is rarely used. Standard practice is to sketch quickly, apply dimensions, and then edit dimensions to precise values. This last steps resizes everything as appropriate.

> **TIPS** *You can drag the menu bar, toolbar, browser, and panel to other locations in the Inventor window. They can float or dock.*
>
> *You can customize the toolbar, keyboard shortcuts, and some other user interface elements through the Tools | Customize command.*

Command Options

Command options are handled by small dialog boxes that appear automatically when needed. Below are examples of these for specifying options, such as dimension lengths, fillet radii, and polygon parameters.

Left to right: Inventor uses dialog boxes in place of AutoCAD's command options.

Mouse Buttons

You'll find that buttons on your mouse operate the same in Inventor as in AutoCAD. The differences are that Inventor doesn't let you change the meaning of buttons, and that it supports just three buttons — against AutoCAD's customizable support for 16 buttons. ("But when's the last time you saw a 16-button mouse?" asks the technical editor.)

Here is how the button definitions compare between the two CAD systems:

Button #	AutoCAD	Inventor*
Left button:		
1	Select	Select
Double-click	Edit properties	Edit part, sketch, or feature
Right button:		
2	Context menu	Context menu
Ctrl+2	Object snaps	Constraints
Roller wheel	Zoom in/out	Zoom out/in, the reverse of AutoCAD**
Click and hold	Pan	Pan
4 - 16	Customizable	Not available

Notes:
*) Inventor does not support the use of tablets as digitizers.
***) When **Reverse Zoom Direction** is enabled in the Tools | Application Options | Display dialog box, the rollerwheel operate zoom in the same manner as AutoCAD.*

3D Mouse

Inventor supports SpaceNavigator, a line of 3D mice from 3Dconnexion. This puck-like mouse lets you move and edit in all six axes: zoom, pan, and rotate — as the illustration provided by 3Dconnexion shows. Inventor on Windows 2000, XP, and Vista is supported through drivers provided by 3Dconnexion.

Typically, you use the SpaceNavigator together with the mouse. You use the mouse with your left hand for object and command selections, while holding the SpaceNavigator in your right hand to control the 3D viewpoint — panning, zooming (pull and push the knob), spinning, and rolling.

Zoom and Pan

Inventor has fewer zoom options than AutoCAD, and that's a good thing. Between documented and undocumented methods, I don't think AutoCAD users need 15 ways to perform zooms. (AutoCAD's Zoom command is rather unique in having no fewer than *four* actions that are defaults!)

Inventor's zooms and pans are straightforward, and their related shortcuts are just as elegant:

Inventor Command	Inventor Shortcut	AutoCAD Equivalent
Previous View	F5	Zoom Previous
Next View*	F6	...
Zoom	*Roller wheel*	Zoom
Zoom All	Home	Zoom All
Zoom Select	End	Zoom Object
Zoom Window	Z	Zoom Window
Look At	PgUp	Plan
Isometric View	...	View -1,-1,1
Rotate	...	3dOrbit
Pan	Drag mouse wheel	Pan

Note:
*) Available only after Previous View.

Press **Esc** to exit zoom, pan, and rotate modes.

ContentCenter (DesignCenter)

Carrying on with our tour of Inventor's user interface....

The Browser has an item called "Favorites" under the Model button. It hides ContentCenter, which is the equivalent of AutoCAD's DesignCenter.

ContentCenter provides drag'n drop access to something like 1.5 million pre-built parts. This is possible, because Inventor uses table-driven part definitions, where a single part file (with embedded spreadsheet) creates *all* threaded fasteners — Imperial and metric — in *all* lengths with *all* head styles, and so on. You can create the equivalent using Inventor's iParts function.

To add parts to drawings, follow these steps:

1. Work your way through the ContentCenter's tree structure to reach the group of parts in which you are interested.

2. Drag the part into the drawing.

3. Notice that Inventor displays a dialog box for choosing a variation of the part.

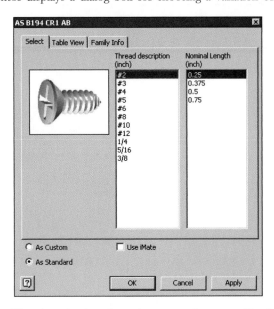

The large number of 1.5 million is arrived at through parametric means. For instance, a single screw, like the one illustrated above, can be placed with eleven different threads-per-inch specifications, each of which has four different lengths. Thus, this one screw part is available in 44 variations.

Further variations are available by editing information in the Family Info tab.

Drawing Sheets (Layouts)

So far you've seen several of Inventor's environments, such as Assembly, Part, Sketch, and Feature. There is one more I'd like to show you, the Drawing environment. It is used for making 2D plans of 3D models on sheets.

This is exactly like layouts in AutoCAD, but Inventor's are much easier to use. As in AutoCAD, Inventor uses viewports to open views into the 3D model. Unlike AutoCAD, Inventor automatically applies hatching, adds detail labels, hides viewport borders, and so on. Inventor makes it so easy that the next 20 pages of tutorials can be completed in two minutes. Seriously.

In this tutorial, you'll create several 2D views of the *analyze-2* model, and then add dimensions, a parts list, and text to the sheet. At the end of the tutorial, the sheet is printed out.

1. From the **New** flyout on the toolbar, choose **Drawing**.

Notice that Inventor opens a new window with something that looks like a drawing sheet, complete with title block and border. I'm not sure why the "paper" is colored pale tan; perhaps it is evocative of aged drafting vellum from an older era.

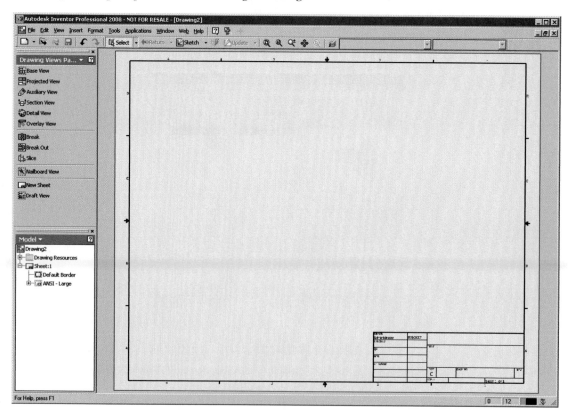

TIP *When you ask the New command to create a new drawing, Inventor doesn't just open a new window; it also creates a new, separate file with extension .idw. This is a major difference from AutoCAD: Inventor separates its "model space" and "paper space" into separate but linked files. (By comparison, AutoCAD combines model and paper space in a single .dwg file.)*

Similarly, every part in every assembly resides in a separate .ipt part file.

2. To add 2D views of the model, follow these steps:

   a. In the **Drawing Views** panel, choose "Base Views."

Notice the dialog box.

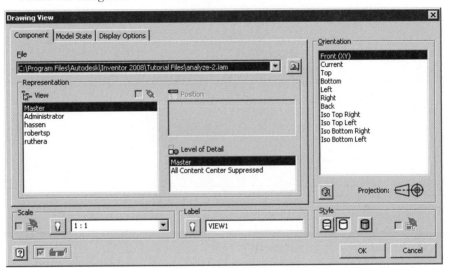

   b. The list you're interested is shown under **Orientation**. It lists the standard views. Choose "Front (XY)," and then click **OK**.

   (You can ignore everything else in this dialog box for now.)

Notice that Inventor places a black-white front view of the model in a viewport. None of that messing around with AutoCAD's SolView, etc. commands!

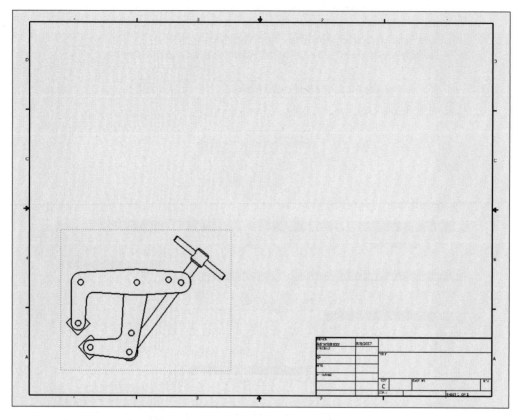

c. If you don't like the location of the view, then drag it to another location on the sheet. Dragging only works, I find, when I pick a point just inside the view's border and the four-arrow icon appears next to the cursor.

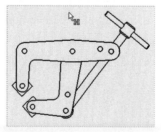

3. Feel free to add projected views, such as top and isometric. In this way, they remain orthogonal to the base view when the parent is moved.

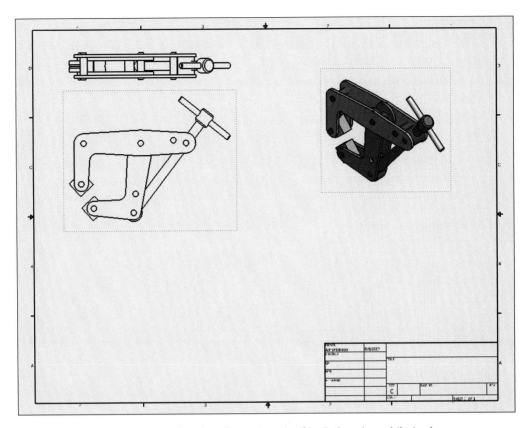

To display the isometric view in color, select the **Shaded** option while in the Drawing View dialog box's Style section. Now that was easy!

So you've gone full-circle: 2D sketches are used to make 3D models, and drawing sheets are used to make 2D representations of 3D models. It is important to remember that in Inventor "drawing" means *2D*. Sheets are drawings with title blocks and drawing borders.

> **TIPS** *Inventor automatically selects the appropriate drawing border based on the drawing standard used for the 3D model — ANSI, in this case.*
>
> *To change the drawing standard, select **Active Standard** from the **Format** menu.*
>
> *To change the drawing border, right click **Layout** in the browser, and then select **Edit Sheet** from the shortcut menu. This allows you to change the border size, such as C to D. New title blocks, revision blocks, and so on can be inserted into the border sheet by accessing them from the **Drawing Resources** folder in the browser.*
>
> *Title blocks can be directly deleted from the sheet by highlighting and then selecting **Delete** from the right click menu.*
>
> *Borders and title blocks can be customized to suit corporate standards; the title block can be set up to fill itself automatically using field text.*

Section Views

It is simple for Inventor to create section, auxiliary, detail, and projected views. If you can draw a line, you can create a section view! Here's how:

1. In the **Drawing Views** panel, click **Section View**.

2. On the status bar, Inventor prompts you:

   Select a view or view sketch *(Pick a view.)*

   Pick one of the views, such as the front view.

3. Inventor prompts you to pick the points that define the section line:

   Enter the endpoints of the section line *(Pick two or more points.)*

   Pick two points to define the location and angle of the section line, such as those illustrated in the figure below. (You would pick more than two points for a segmented section view.)

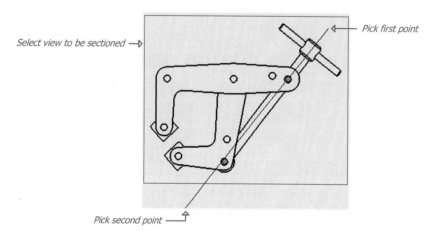

Select view to be sectioned →

Pick first point

Pick second point —

4. When done picking points, right-click and choose **Continue** from the shortcut menu, as prompted by Inventor on the status bar:

   Enter additional points or choose "Continue" from the right-click menu

Inventor previews the section that'll be generated, and opens up the Section View dialog box. This allows you to define the numbering system and other properties of the section.

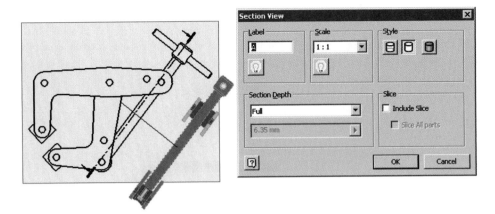

5. Follow Inventor's prompt to pick a point to locate the section view. If necessary, you can relocate it later.

   Select view location *(Drag the section away, and then pick a point to position.)*

6. Click **OK** to exit the dialog box. The section appears fully hatched in the drawing, along with the section's "A-A" label. As I said, it's as easy as drawing a line.

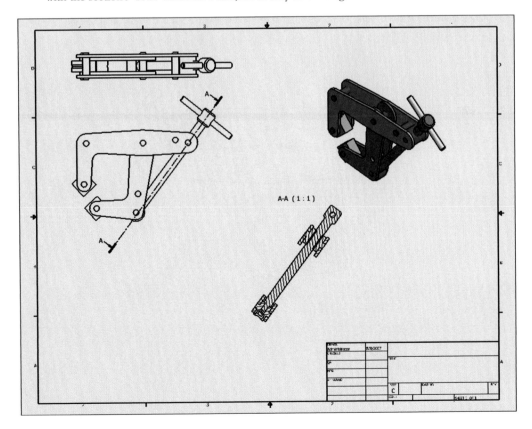

Retrieving Dimensions

Inventor has two classes of dimension. The first class you saw earlier in this chapter — driven dimensions that define the sizes of sketches.

The second class is more like the dimensions you are used to in AutoCAD: you access Inventor's sketch and model dimensions in the Drawing environment. These dimensions you can draw (as in AutoCAD) or *retrieve*. When you retrieve dimensions, you convert the model's driven dimensions into conventional ones.

Here's how to retrieve dimensions:

1. Select a view, and then right-click.

2. From the shortcut menu, select **Retrieve Dimensions**. Notice the Retrieve Dimensions dialog box.

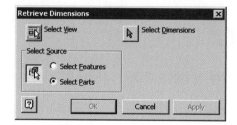

3. Click **Select View**, and then choose a viewport, such as the one showing the front view.

4. Click **Select Dimensions**. Notice that Inventor displays *all* the driven dimensions attached to the model.

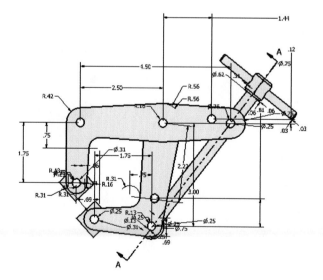

5. Such a large number of dimensions is overwhelming, and so you'll just choose a few.

   With the cursor, select several dimensions, such as the ones highlighted in the figure below. As you do, notice that Inventor highlights them with green dots and blue lines.

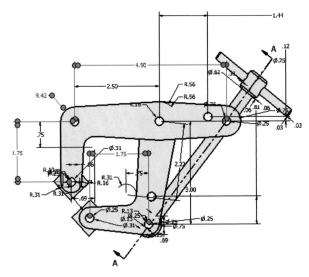

6. In the Retrieve Dimensions dialog box, click **OK**. Notice that the selected dimensions remain, while the unselected ones disappear.

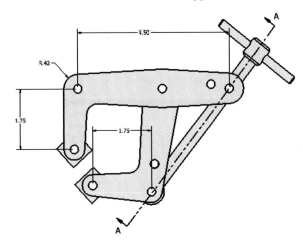

7. You're almost done. There is just the problem of the horizontal dimension reading 1.75: it is misplaced. It shouldn't cross any of the parts.

To fix this problem, grab the dimension line (anywhere near the dimension text) and then drag it below the parts.

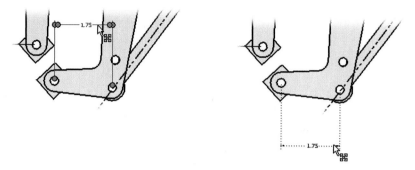

Left: *Selecting the dimension.*
Right: *Dragging the dimension line downwards.*

TIPS *When dimensions are selected, grips appear for editing them. The purposes of the grips are illustrated below:*

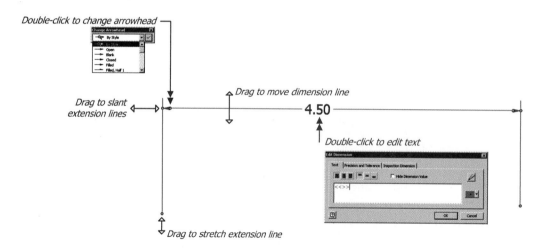

Right-click the dimension for additional editing options, such as toggling the display of individual arrowheads and choosing other dimension styles.

To draw dimensions manually, switch the panel to Drawing Annotation panel, and then choose from the dimension types available, such as General (linear). baseline, and ordinate. Like AutoCAD's DimLinear command, Inventor's General Dimension draws horizontal, vertical, or aligned dimensions, depending on where you move the cursor.

But, notes the technical editor, here is the real power: right-click a dimension, and then choose **Edit Model Dimension** *from the shortcut menu. When you change the value of the dimension, the part resizes itself in all drawing views, in every part file, and in the assembly model.*

To delete dimensions, select them, and then press the **Del** *key.*

Generating Parts Lists (DataExtraction & Table)

Inventor generates parts lists more easily than does AutoCAD 2008 with its DataExtraction-to-table method. Here's how to go about it:

1. In the panel's droplist, click **Drawing Views Panel**, and then choose "Drawing Annotation Panel."

2. Click **Parts List**.

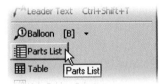

3. Notice the Parts List dialog box. Select a view from which to extract the parts list.

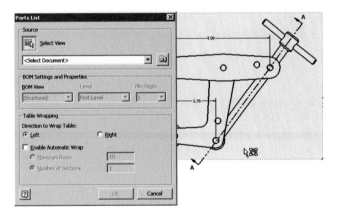

4. Click **OK**.

5. Notice the rectangle. It outlines the parts list table. Drag it into position.

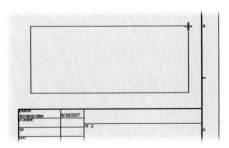

6. Click the left mouse button, and the parts table appears.

ITEM	QTY	PART NUMBER	DESCRIPTION
		Parts List	
1	2	Upper Plate	
2	2	Lower_Plate	
3	4	Pin_A	
4	1	Pin_B	
5	1	Pivot_Threaded	
6	1	Handle	
7	1	Screw	
8	1	Pivot_Lower	
9	1	SHCS_10-32x6	
10	2	ch_09-Grip	

The green handles resize the table.

Right-click the table for additional options, such as rotating it and exporting its data. To edit the content, double-click the table.

Once placed, the table can be dragged to another location. You can also copy and paste (**Ctrl+C** and **Ctrl+V**) the table into other drawings.

Placing Text

Let's do one more thing before you're ready to plot the drawing: place text in the title block. Assuming Inventor is still showing its Drawing Annotation panel, here's how to do it:

1. Use the "Z" shortcut to do a windowed zoom into the title block area: at the 'Ready' prompt, enter **z**.

   Ready z
   Zoom into a window (Pick a point, and then pick a second point to specify the size of the zoom rectangle.)

   Notice that Inventor's green rectangle previews the actual ratio of the new view size, unlike AutoCAD. (See figure on the next page.) That's why the rectangle's corners don't necessarily match your pick points.

2. On the panel, click **Text**.(Alternatively, enter the "T" shortcut at the keyboard.)

3. Notice that Inventor prompts you on the status bar:

   Click on a location or two corners

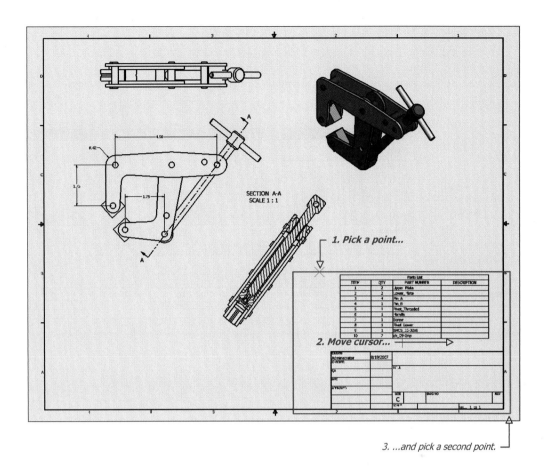

1. Pick a point...

2. Move cursor...

3. ...and pick a second point.

Single-point Text

AutoCAD needs three commands to place text — Text, DText, and MText — but Inventor needs just one. The Text command places all text as multi-line text (mtext) — even though its prompt seems to imply you can place single or multi line text with its "Click on a location or two corners" prompt. Whether you click one point or pick two points, the text is effectively multi-line in nature. To show the minor differences between the two, you'll use both methods in this tutorial.

Notice that Inventor has already filled out some of the title block with the name of the drafter (DRAWN) and today's date. These are placed using field text, just as in AutoCAD.

For this tutorial, you'll first add the revision number in the REV box in the lower right corner.

1. In the REV box, move the cursor around. Notice the yellow grip that jumps around the box. Inventor is showing you the snap points.

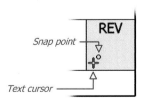

2. Pick a point that approximates the lower-left corner of where the text should begin. This is similar to AutoCAD's default Left justification.

   Notice the dialog box. It's somewhat like that of AutoCAD's MText command with options on top and a text entry area below. But instead of tabs and shortcut menus creating a rather compact layout in AutoCAD, Inventor's Format Text dialog box lays everything out in the open.

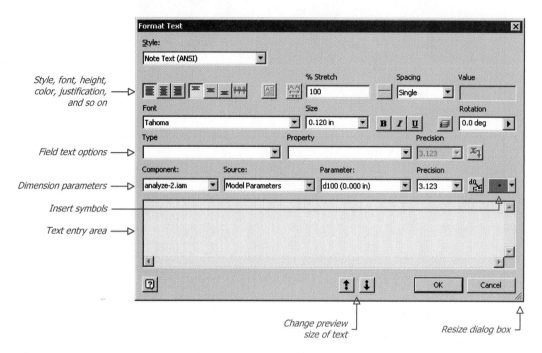

3. In the text preview area, enter "A."

4. Notice that the vertical justification is set to top. That means the text will be situated too low. Click the **Bottom Justification** button.

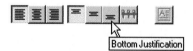

5. Click **OK**.

   Notice that the text is placed in the REV box.

Exiting the dialog box does not exit the Text command; it continues running until you press **Esc**. For now, though, keep it running so you can place more text.

Rectangle Text (MText)

This time you'll place text by specifying a rectangle through picking two corners. I find it a bit annoying that this method differs from that of Zoom Window.

1. Instead of picking two points, you do the following:

   a. Pick one point with the left mouse button.

   b. Keeping the mouse button held down, drag the cursor to show the size of the rectangle.

   c. Let go of the mouse button.

   The figure below illustrates the procedure. This is equivalent to creating an mtext boundary in AutoCAD.

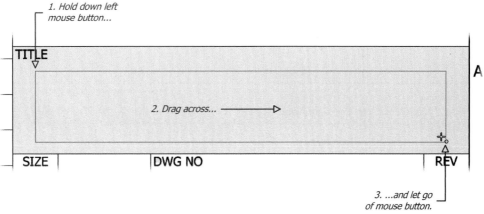

2. After letting go of the mouse button, notice that the Text Format dialog box reappears.

3. For this tutorial, you'll add the drawing's file name to the title area as *field* text. Follow these steps:

   a. From the **Type** droplist, choose "Properties - Model."

   b. From the **Property** droplist, choose "FILENAME."

   c. To add the text, click the ⊠ **Add Text Parameter** button.

   Notice that the field text is shown in red and in angle brackets in the preview area.

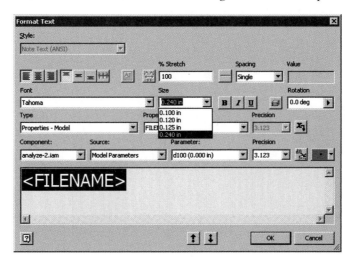

4. The preview area shows the text at its actual size. As a title, it should be larger, like this:

   a. Select the "<FILENAME>" text.

   b. Choose "0.240" from the **Size** droplist.

   Notice that the text doubles in size.

5. Click **OK** to exit the dialog box and place the text.

DRAWN Administrator	8/19/2007			
CHECKED				
QA		TITLE **analyze-2.iam**		A
MFG				
APPROVED				
		SIZE **C**	DWG NO	REV **A**
		SCALE	SHEET 1 OF 1	

      2   1

6. Press **Esc** to exit the Text command.

When text is selected, it has eight grips for resizing its bounding rectangle.

To edit the text, double-click it. (To see the entire drawing, press the **Home** key.)

Plotting Drawings

The drawing is done, so let's plot it. Contrasted with AutoCAD's myriad of options, Inventor's is about as basic as it gets: choose a printer, and then print.

It's this simple for two reasons:

- Inventor takes care of **scaling** in a way that AutoCAD doesn't. Did you notice during the tutorials that you never once had to set a scale factor? Inventor handles this for you when you chose the paper size in the Drawing environment — a paper size that matches that of the printer or plotter. Once Inventor knows the paper size, it automatically determines the scale factor for all scale dependent objects, such as hatches, linetypes, and text height.

- Inventor drawings are just as often published paper-free as 2D and 3D DWF files or in other **electronic** formats. In this case, paper size doesn't matter.

Print Setup

With that brief introduction behind you, prepare to plot the drawing:

1. From the **File** menu, choose **Print Setup**. Notice the rather sparse Print Setup dialog box.

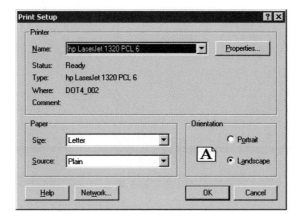

2. In the dialog box, ensure that **Orientation** is set to "Landscape."

   Confirm that the correct printer is selected. (If necessary, click **Properties** to set the resolution, print quality, and so on.)

3. Click **OK**.

Note that Inventor's options for plotting don't show up until you later execute the Plot command. Inventor has no AutoCAD-like notion of plot styles.

Print Preview

To preview the drawing before committing it to paper, follow these steps:

1. From the **File** menu, choose **Print Preview**. Notice the preview window.

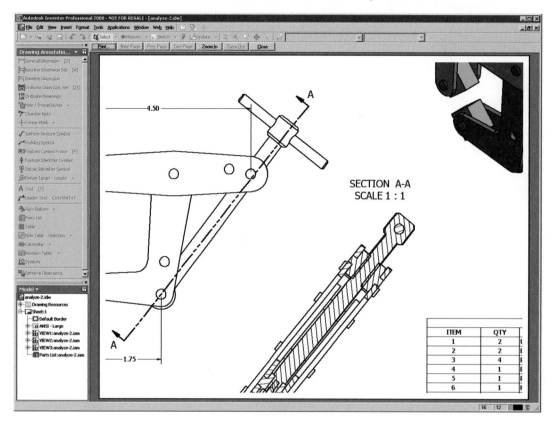

Oops! The drawing is too big. Now what? *Ensure the sheet size matches the printer!* You have two choices to fix this problem:

- Change the sheet size to match the printer paper, which is A-size in this case. This requires that the views be scaled downward, such as by 1:2.

- Have Inventor plot the drawing to fit the paper; the drawing will probably not be plotted at a standard scale.

- Use any border format with the same aspect ratio as A-size, such as C.

I describe the first two choices below.

2. Click **Close** to return to the drawing.

Appropriately Sized Sheet

The problem of the "too large" drawing stems from Inventor selecting the default drawing size as "ANSI-Large." A solution is to change the sheet to an A-size — the drawback being that all the views need to be reestablished.

Here's how:

1. In the Model panel, open **Sheet Formats**.

2. Right-click "A size, Landscape, 1 view," and then choose **New Sheet** from the shortcut menu. ("1 view" means that one view will be automatically created when the new sheet is created.)

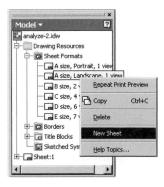

3. Inventor displays a dialog box for selecting the model to be viewed here.

Ensure "analyze-2.iam" is chosen, and then click **OK**.

Whoa — same problem! The view is too large.

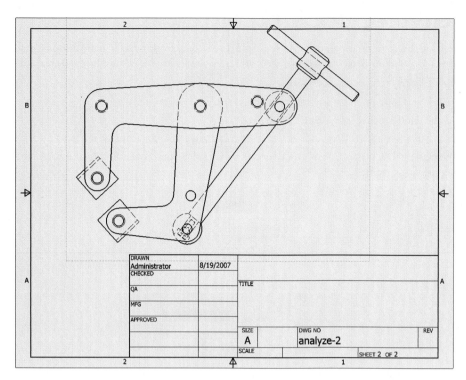

This occurs because Inventor assumed a view scale factor of 1:1. The solution is to change the scale, as follows:

a. Right-click the base view.

b. From the shortcut menu, choose **Edit View**.

Notice the familiar Drawing View dialog box.

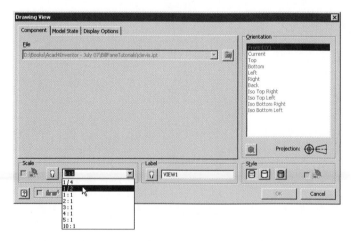

c. From the **Scale** droplist, select "1/2," and then click **OK**.

Notice that base view becomes half as large. In addition, all child views also change their scale to match — but text and dimension heads stay the same size.

(Alternatively, you can set the scale factor in the Drawing View dialog box.)

4. Use Print Preview to ensure the drawing fits the paper, and then click **Print** to plot it.

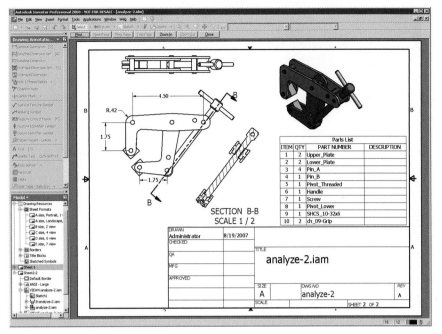

Scaling the Drawing

The second solution is to have Inventor fit the drawing sheet to the paper. This is done after you execute the Print command, as follows:

1. From the **File** menu, choose **Print**. Notice the Print Drawing dialog box. Personally, I think this dialog box ought to appear with the Print Setup command, but oh well.

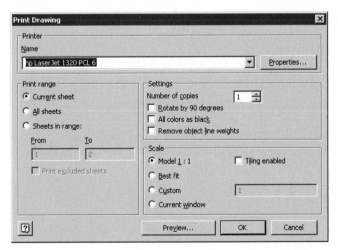

2. Here you have several choices for selecting sheets, scaling, and setting all colors to black.

   One of the choices is **Best Fit**, which fits the sheet to the paper, scaling it as required.

3. Click **OK**, and then wait a few moments for the drawing to appear from your printer.

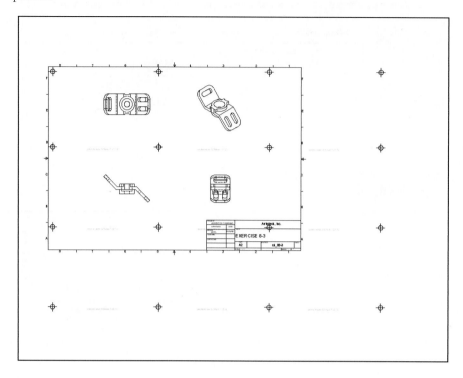

TIP *Inventor provides an alternative to scaling large sheets to fit small sizes of paper. Turn on the **Tiling Enabled** option to print sheets full-size on four or more pieces of paper.*

The pain comes after, when you cut and tape the pages together to make the single sheet. (Print preview uses ⊕ *icons to indicate the corners at which the pieces of paper meet, as illustrated above.)*

Summary

This concludes the introductory tutorial, giving you wet feet by wading into some of Inventor's capabilities.

The next chapter describes differences and similarities between Inventor and AutoCAD in great detail, while the chapter following provides a step-by-step tutorial on drawing 3D models with Inventor.

Chapter 3

Inventor & AutoCAD,
Similarities & Differences

When you first start Inventor, you'll find a user interface that looks vaguely familiar. See the figure spread across the next two pages. In this chapter, I'll tell you about the similarities and differences between AutoCAD and Inventor.

The reason for the similarities is deliberate. This is because Autodesk has been reworking AutoCAD to make it look more like Inventor, specifically through the AutoCAD workspace named "3D Modeling." The reason for the differences is that Inventor was initially a secret project, developed independently of AutoCAD.

In AutoCAD, the most visible changes were adding the Dashboard, displaying the grid with lines, changing from ShadeMode to visual styles, adding the ability to draw and edit in perspective viewing

IN THIS CHAPTER

- Touring the Inventor user interface.
- Starting new drawings.
- Learning how AutoCAD DWG files act in Inventor, and vice versa.
- Understanding the four environments and their four files.
- Making selections, changing properties, and seelcting drawing standards.
- Using snaps, grid, units, lineweights, materials, layers (kind of), osnap, limits, elevation, thickness, hatches, visual styles, shadows, arcs , lines, dimensions and constraints.
- Editing lines, circles, arcs, text and styles, attributes and BOMs.
- Plotting and publishing 3D models.

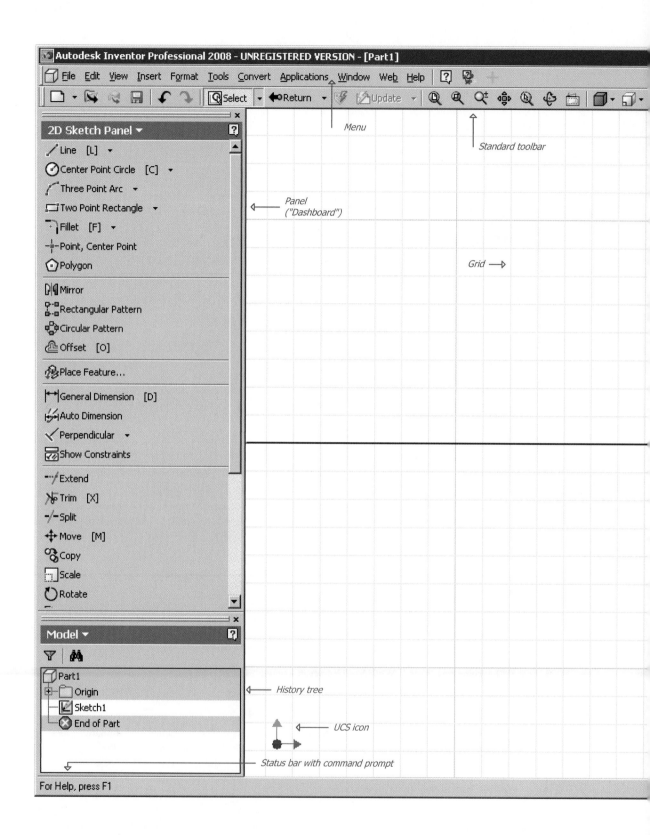

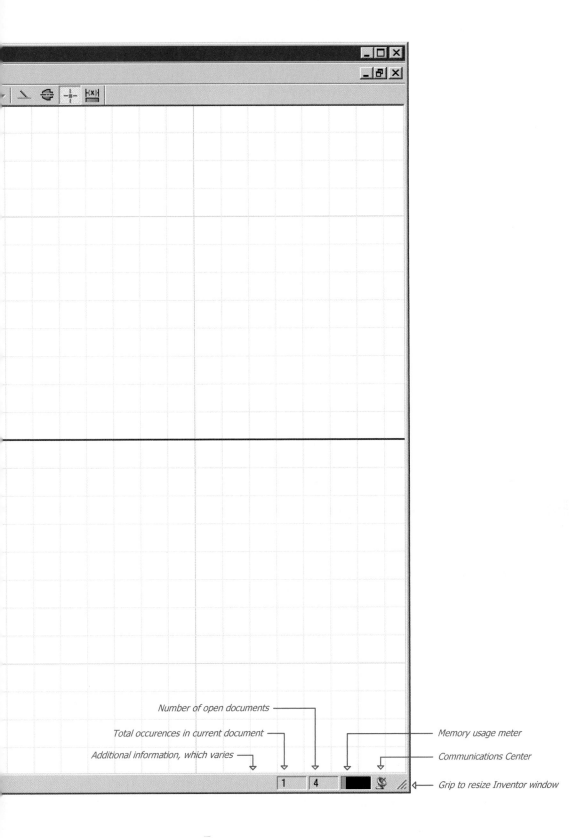

Number of open documents

Total occurences in current document

Additional information, which varies

Memory usage meter

Communications Center

Grip to resize Inventor window

mode, including Inventor-friendly commands, such as DimInspect, and reducing of the number of toolbars.

Under the hood, AutoCAD had been using the same solid modeling kernel (ShapeManager) as Inventor. In AutoCAD 2007, Autodesk switched its rendering engine to the same one used by Inventor, 'mental ray.'

Also added to make AutoCAD more Inventor-compatible were true surfacing, easier face selection, automatic UCS orientation, realtime editing of 3D models, and a simple version of history.

At the same time, Autodesk was tweaking Inventor 2008 to make it friendlier towards AutoCAD, for example opening Inventor *.dwg* drawing files in AutoCAD without a separate translation step.

(All of which makes me wonder, "Why Inventor?" I find it interesting that Autodesk isn't working to make AutoCAD more like Revit, Autodesk's 3D modeling software for architects.)

Touring the Inventor User Interface

Start Inventor by double-clicking its icon on the Windows desktop.

Autodesk Inventor
Professional 2008

Notice the Open dialog box.

Here you have the choice of starting several kinds of documents or models:

- One or more documents native to Inventor. You can open two or more files at once by holding down the **Ctrl** key as you select their file names.

- Other kinds of files, such as AutoCAD *.dwg* drawings. You change the file type with the **Files of Type** droplist.

- New drawings. Click the **Begin New Drawing** button in the Quick Launch area.

New Documents

In the last chapter, you opened existing documents in Inventor. Inventor can start new documents in one of several drawing modes, which you'll learn about later in this chapter. The basic ones are Part, Assembly, Drawing, and Presentation, but there are also environments for specialized applications, such as electrical wiring and sheet metal design.

To have Inventor open a new (blank) document, continue with these steps:

1. Click the ▭ **New** button in the Open dialog box (or the toolbar, if the dialog box is closed).

2. In the New File dialog box's **Default** tab, choose "Standard.ipt," the template for new parts.

   The *.ipt* file extension is short for Inventor Parts Template.

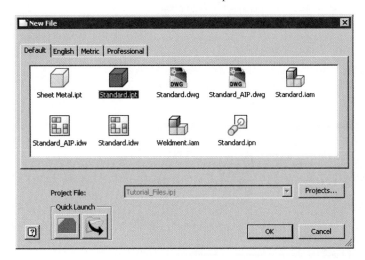

3. Click **OK**. (Alternatively, double-click the template icon.)

Notice that Inventor opens in the Parts environment, as illustrated on the previous pages.

> **TIPS** *"Project files" specify the locations where Inventor stores all files related to the named project.*
> *"Environments" are the equivalent of AutoCAD's workspaces.*

FINDING SAMPLE DRAWINGS

Inventor includes dozens of sample drawings that new users may have difficulty locating. The solution is to use the Open dialog box's **Projects** droplist. *Project files* store paths to folders containing Inventor files specific to individual design projects; Autodesk uses project files *(.ipj)* to point to samples, tutorials, and other files.

You cannot change projects when files are open, so you have to first close any open files. Then, in the Open dialog box, click the **Projects** droplist. From the droplist, choose a project file name, such as "Samples.ipj."

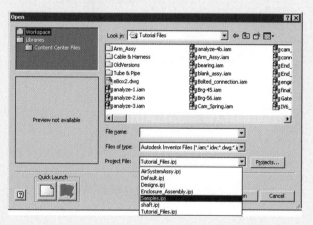

Suddenly, a dozen folders appear, containing scores of assemblies, parts, and drawings for your enjoyment.

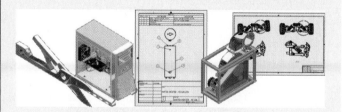

The figure at right list the folder structure used by Inventor. In particular, notice the long list of folders that contain sample drawings under "Samples."

```
Desktop
  Documents
  Computer
    This (C:)
      Documents and Settings
      epson
      Graphics
      Internet
      NVIDIA
      Program Files
        AnswerWorks 4.0
        AOEMView 2008
        Autodesk
          Autodesk DWF Writer
          Data Management Server 2008
          DWG TrueView
          Inventor 2008
            Backgrounds
            Bin
            Catalog
            Compatibility
            Configuration
            Design Accelerator
            Design Data
            LiveUpdate
            Preferences
            PSS
            Samples
              Content Center Files
              IDFTranslator
              Models
                AEC Exchange
                Assemblies
                  Arbor Press
                  Blower
                  Engine MKII
                  Metal Container
                  Personal Computer
                  Scissors
                  Shaver
                  Stapler
                  Suspension
                  Suspension Fork
                  Test Station
                  Tuner
                Cable & Harness
                  Rackmount Enclosure
                  Report Generator
                  Wire Library
                  Wire List Import
                Dynamic Simulation
                  Gate
                  Windshield Wiper
                Frame Generator
                  Camper Frame Weldment
                Parts
                  Flower
                  Hairdryer
                  OilPan
                  Plate
                  Pump Housing
                  Speedometer
                  TireRim
                Sheet Metal
                  Electrical Box
                  Mounting Bracket
                Stress Analysis
                  Bracket
                Translation
                  Arm Rest
                Tube & Pipe
                  Accumulator
                  Cooling Tower
                  Tank
                Weldments
                  Carriage
                  Cosmetic
            SDK
            Setup
            Stress Analysis
            Templates
            Textures
            Tutorial Files
              Arm_Assy
              Cable & Harness
              OldVersions
              Tube & Pipe
            Web
            WebServices
        Patches
      Vault 2008
```

About DWG Files

As an AutoCAD user, you live and die by *.dwg* files, and you may wonder about using AutoCAD drawings in Inventor.

The good news is that Inventor reads and writes AutoCAD drawings; the bad news is that there are two variants of *.dwg* file, and they're not as compatible as you might prefer. There is the DWG format created by AutoCAD and a second created by Inventor — yet they are not 100% interchangeable, because Inventor adds custom objects that AutoCAD can display and edit somewhat, but not fully.

Each CAD program reads the *.dwg* files created by the other; the issue is how much editing is possible — something that is covered in great detail later in this book.

To create a *.dwg*-based drawing in Inventor, you have start out with one: choose "Standard.dwg" from the New File dialog box's list of templates. This action opens Inventor's Drawing environment.

How AutoCAD DWG Files Act in Inventor

Inventor directly opens *.dwg* files created by AutoCAD, but then is limited to viewing, measuring, and plotting them. This means Inventor can be used to view *.dwg* files without AutoCAD.

You can use the following workaround to get around the view-only limitation: start a new part file, and then open the *.dwg* file in Inventor. Copy objects to the Clipboard from the open *.dwg* file, and then paste them as 2D sketch objects in Inventor's Sketch environment of the new part file. From there, you convert the sketches into Inventor 3D objects.

Objects created in AutoCAD are read-only AutoCAD objects in Inventor – with a few exceptions: Blocks, layers, and layouts can edited in both AutoCAD and Inventor.

Text and dimension styles are synchronized: changes made in one CAD system are applied to the other. AutoCAD's layouts are displayed as sheets in Inventor. You can create views and place annotations on AutoCAD layouts in Inventor; the added data coexists with the AutoCAD data.

Using AutoCAD as Templates for Inventor

Inventor users and Inventor's documentation talk about using "AutoCAD templates" with Inventor. As an AutoCAD user, you know that means working with *.dwt* template files. But Inventor users mean something different: opening regular *.dwg* drawing files and then using the content therein with Inventor. The idea here is that you can create title blocks and drawing borders in AutoCAD's layout mode, and then use them in Inventor. This lets you reuse your company's existing borders and title blocks.

(You would think it would be a simple thing for Inventor to open *.dwt* files, but no, you have to use Windows Explorer manually to change the extension to *.dwg*.)

You can create new Inventor drawings from DWG files, but all graphical AutoCAD data will be removed — other than block instances in layouts (blocks in model space are erased). Turn anything you want to keep into blocks. All non-graphical data is preserved, including dimension and text styles and layers.

How Inventor DWG Files Act in AutoCAD

Inventor also creates and saves files in DWG format. The data, however, are saved in Inventor's version of the format. The drawings look the same in both Inventor and AutoCAD. In AutoCAD, you can view, measure, and plot the Inventor data.

You can explode and delete Inventor objects while in AutoCAD. But this is strongly discouraged, because the intent is that all editing be performed in Inventor, with the changes reflects in the AutoCAD file.

Inventor creates AutoCAD-compatible block definitions for every drawing view and sketch in its files. (As you edit them in Inventor, the block definitions are updated automatically.) When the Inventor-format *.dwg* file is opened in AutoCAD, you can use the Insert and ADCenter commands to place the blocks, or use the XRef command to reference them in other drawings. This creates *hybrid* drawings, such as where Inventor revises portions of drawings originally designed in AutoCAD.

The figure below shows a drawing in Inventor...

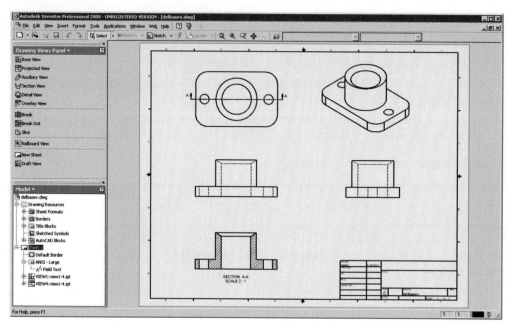

...and the same drawing opened in AutoCAD.

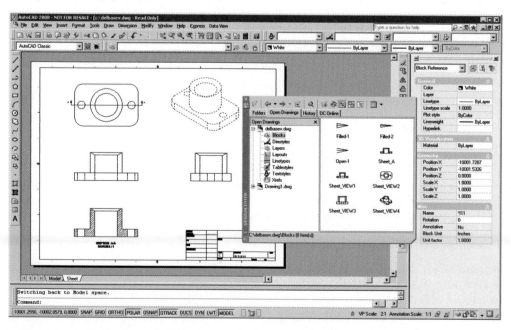

In the figure above, DesignCenter is displaying preview images of the Inventor-generated blocks (views), while the Properties palette lists information about one of the selected view-blocks.

Using the Recover command on Inventor *.dwg* files in AutoCAD 2007 (and earlier) removes some Inventor data.

Four Environments, Four Files

Perhaps the most challenging thing for me initially to figure out was that Inventor has four environments: Part, Assembly, Drawing, and Presentation. When you create 3D models in Inventor, you generally follow this path, moving from one environment to the next:

1. Part.

2. Assembly.

3. Presentation.

4. Drawing.

Environments in Inventor are like workspaces in AutoCAD.

I lied. There are many more than just four environments. But these are the basic four listed by the New button on Inventor's toolbar.

The others are variations on these ones, such as the Sketch "sub" environment found in the Part, Assembly, and Drawing environments.

Part (*.ipt* files) is the environment for creating 2D sketches (a.k.a. *profiles*), and then adding 3D features to make parts. This is equivalent to AutoCAD's Model tab. (When AutoCAD 2D drawings are imported, they are treated as sketches.)

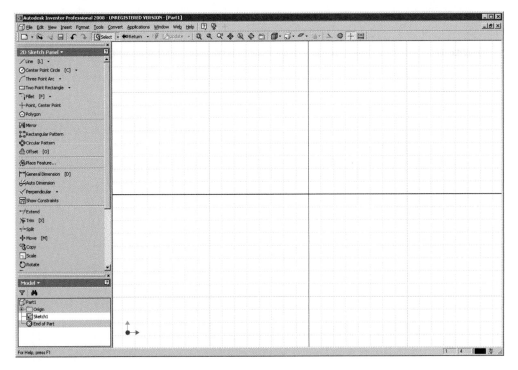

Assembly (*.iam* files) is the environment for collection parts (*components*) in assemblies. There are a few 3D editing commands, such as Extrude and Hole, but when you edit a part, Inventor switches back to its Part environment, where you regain access to sketching and editing commands; when done, click the **Return** button to return to the Assembly environment.

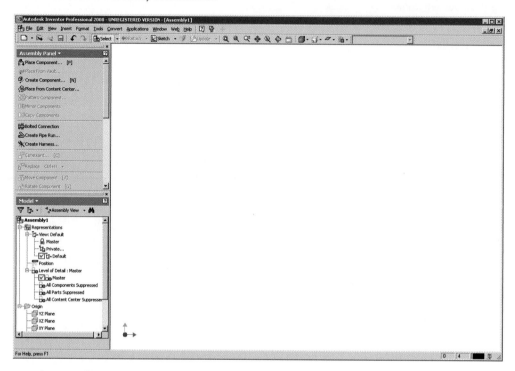

Presentation (*.ipn* files) is the environment for creating exploded views, animations, and stylized views of assemblies. Models cannot be edited in this environment.

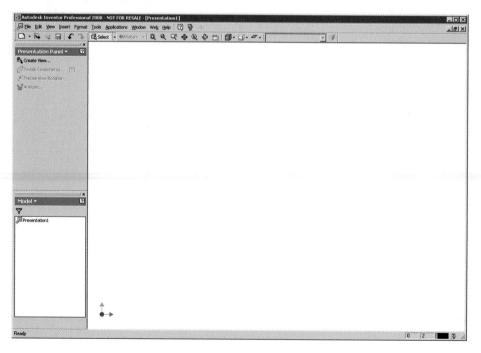

Drawing (*.idw* files) is the environment for creating and annotating 2D and 3D views of parts and assemblies, as well as placing drawing borders and parts lists. This is equivalent to AutoCAD's Layout tab.

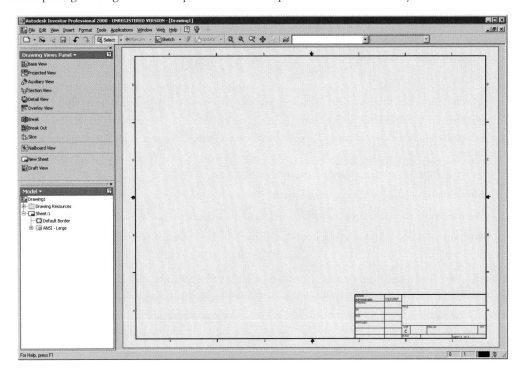

Each of these four is stored in a separate file. Even though everything is stored in separate files, the files interconnect. Make a change to a part, and its representation is updated in the assembly, drawing, and presentation files.

Project (*.ipj* files) contain pointers to paths so that Inventor can find the files it needs. Parts, assemblies, and so on are not stored within project files.

When I think about it, AutoCAD also has four drawings "environments:" model tab, layout tabs, model space viewports in layout tabs, and the block editor.

The following table is an approximate correlation between the drawing environments of AutoCAD and Inventor:

Inventor	AutoCAD
Part	Model tab
Assembly	Sheetsets
Presentation	Rendering in model tab
Drawing	Layout tab

Additional Environments

In addition to the four described above, Inventor has many more drawing environments, as follows:

iFeature (*.ide* files) — creating and editing multiple versions of parts, kind of like AutoCAD's Block Editor.

Construction — stitching imported surfaces into whole surfaces and solids.

Solids — modifying solids.

3D Sketch — creating paths for 3D sweep features that route wiring, cabling, and tubing in parts and assemblies.

Nailboard (*.idb* files) — creating 2D flattened views of wire harnesses, cables, and ribbon cables.

Tube & Pipe — routing pipes, tubes, and hoses.

Sheet Metal — designing sheet metal parts.

Weldment — accessing tools specific to welding and assembly.

Stress Analysis — simulating the behavior of parts under loads and frequencies.

...and there's more! If you are curious about the names of all 31 environments, click **Tools | Customize**, choose the **Environments** tab, and then click the **Environment** droplist.

In addition, Inventor also has a plot preview mode, and a separate rendering and animation environment, called "Studio." Whereas AutoCAD only provides camera animation, Inventor allows both the camera and the parts to move.

AutoCAD's Similarities to Inventor

Because of the changes Autodesk made to AutoCAD 2007 and its 3D Modeling workspace, many aspects of Inventor's user interface should look mildly familiar to you. There are the menus, the toolbars, the "Dashboard," and so on.

Whereas AutoCAD uses "workspaces," Inventor uses "environments" to control which toolbars, menus, keyboard shortcuts, and so on, are available.

While AutoCAD is overwhelmingly customizable, Inventor is less so. If you are used to customizing AutoCAD to the hilt, then you'll be heartened to know that the following items are waiting for you to customize them in Inventor:

Linetypes — use the same format as AutoCAD's *.lin* files.

User interface colors — accessed through Tools | Application Options, and then choose the Colors tab.

Inventor's operation — accessed through Tools | Application Options, then choose any tab

Current document — accessed through Tools | Document Settings.

Toolbars — accessed through Tools | Customize, and then choose Toolbars tab.

Shortcut keystrokes — accessed through Tools | Customize, and then Keystrokes tab. See the dialog box illustrated below.

Styles and Standards — accessed through Format | Styles and Standard Editor.

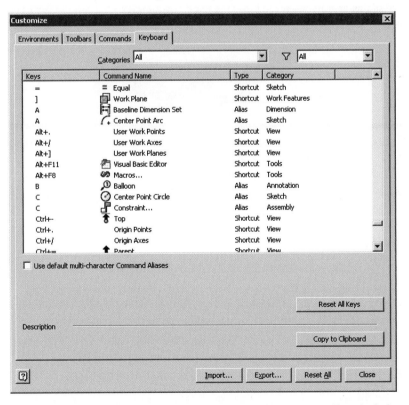

There are things you thought you might be able to customize in Inventor, but cannot. These include panels (dashboard), menus, colorbooks, and the status bar. The technical editor reports that he's never felt the need to change them.

Making Selections

As in AutoCAD, Inventor's default command is **Select**. On the surface, Inventor appears to have fewer methods than AutoCAD for selecting objects. There's pick, select all, crossing, and window.

But in 3D mode, Inventor becomes more sophisticated. Let's start by taking a look at the Select button on toolbar and its options

Select Groups — selects groups of parts.

Feature Priority — selects work planes, axes, and points first.

Select Faces and Edges — selects only faces and edges.

Select Sketch Features — selects features.

Select Wires — selects wires; available in the Wiring Harness module.

In AutoCAD, you would hold down the **Ctrl** key to select conflicting faces and edges. Inventor instead has a little green tool that appears when you hover the cursor over overlapping features for 1 second. (The time interval can be changed by you.)

As you click the arrows at either end of the icon, Inventor highlights different faces (loops), edges, or features. Inventor tends to "tunnel down." This means that it begins with the object closest to your view; as you click the arrows, it moves through all overlapping objects until it gets to the one farthest away.

Once Inventor has highlighted the feature you want, click the green button in the center to make the selection.

Select the object a second time to unselect it.

Changing Properties

Whereas AutoCAD has extensive properties for every object, Inventor has just these few:

Color — select from the Windows dialog box of 48 colors, or specify by HSL and RGB. No ByLayer, ByBlock, or ByEntity settings; no color books or ACI. Default is like AutoCAD's color 7: changes color, depending on the background color.

Linetype — choose from a list of pre-loaded linetypes (unlike AutoCAD); you can import *.lin* linetype definition files, but complex linetypes are not supported.

Lineweight — choose from fixed list; same as AutoCAD's lineweights.

Linetype scale — specify any value larger than zero.

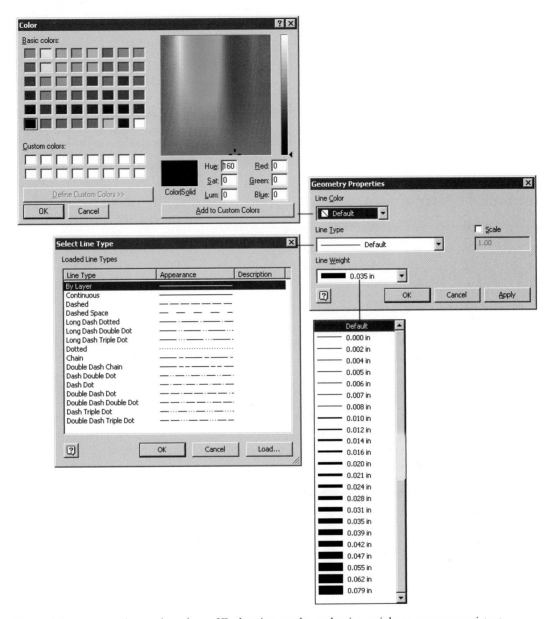

Most of these properties apply only to 2D drawing mode, and exist mainly to ensure consistent appearance between AutoCAD and Inventor drawings.

AutoCAD's Properties palette can stay open always, but the equivalent in Inventor is a dialog box and thus needs to be closed before you continue working on the drawing.

Drawing Standards

Inventor includes the following drawings standards:

> **ANSI** — United States.
>
> **BSI** — British.
>
> **DIN** — German.

GB — Chinese.

ISO — International.

JIS — Japanese.

AutoCAD used to have all the same standards (except for British), but as of 2008, AutoCAD includes only ANSI and ISO.

Inventor doesn't have AutoCAD's Standards and CheckStandards commands. Instead, standards are specified by template files. You can import and export styles through *.styxml* files.

Inventor's Similarities to AutoCAD

Inventor 2008 has some features you're familiar with in AutoCAD:

- Press the spacebar to end and repeat commands.
- Use multikey shortcut keystrokes to activate command (previous releases of Inventor already used single-key shortcuts).
- The tri-color UCS icon cannot be customized.

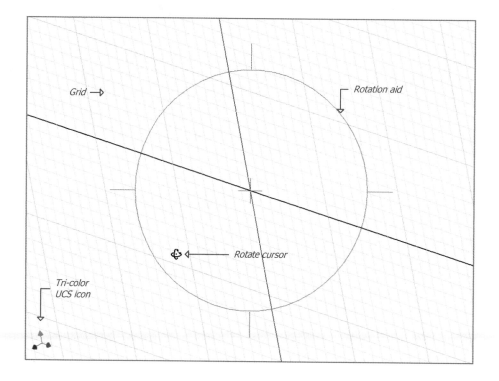

Zoom and pan are essentially identical in both CAD packages, except that there are fewer ways to zoom in Inventor. Unique is the Look At command, which rotates the model so that the selected object is oriented correctly in the viewport — kind of UCS in reverse. (Instead of the UCS rotating to match the feature, the feature rotates to match the WCS.)

The Rotate command is equivalent to AutoCAD's 3dOrbit command, as illustrated above.

Inventor has paper space-like viewports in its Drawing environment. The Presentation environment can have alternative configurations — exploded, assembled, and so on — but only one active at a time. There are no model-space style viewports; the closest equivalent would be the **Window | Arrange All** command.

You can customize toolbars, panels, and shortcut keystrokes. The changes can be shared with other Inventor users through .*xml* files. (AutoCAD does the same through its XML-format .*cui* file.)

For what it's worth, Inventor lacks system variables.

Significant Differences

Because of its origins in the early days of personal computing — before the mouse, when the keyboard ruled supreme (although there were digitizing tablets) — AutoCAD still expects you to enter many command options at the command prompt, or its newer replacement dynamic input (at the cursor).

First developed 15 years after AutoCAD — in the age of Windows and the mouse — Inventor lacks the keyboard-oriented legacy. It does, however, display prompt statements at the left end of the status bar.

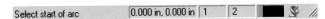

When drawing a three-point arc, for instance, you will see the following prompts in succession:

Select start of arc: *(Pick a point.)*
Select end of arc: *(Pick another point.)*
Select point on arc: *(Pick a third point.)*

But you cannot respond to prompts at the keyboard, such as entering x, y coordinates. Neither can you specify the arc's radius or its chord or angle, nor specify object snaps or point filters. No, no, no. All you need to do is pick points in the drawing, and Inventor figures out what you most likely want. Pick close enough to another object, and Inventor "jumps" the cursor, almost as if it were using object snaps from AutoCAD.

The closest Inventor gets to something like osnaps is constraints. If osnaps are like Lego — objects snapped together geometrically but which come apart when you pull at them — then constraints are more like using glue, where one part moves with the other.

The pick points are not exactly "random." Inventor uses a default snap distance of 0.063 in (1/16th-inch, or 0.1mm in metric drawings) to assign x,y coordinates to objects. This value can be set by the user (Tools | Document settings | Sketch) for the current file, which can then be saved as a template file.

Snaps, Grid, Units, Lineweights, Materials

Inventor's Tools | Document Settings command provides access to many settings similar to AutoCAD's DSettings command. These are per-document settings, meaning you can have different settings for every set of documents. In addition, there are different settings for the Parts, Drawing, Assemblies, and other environments.

Let's take a look at some of the settings in this dialog box, and compare them with AutoCAD:

Snap — snap distance can be specified in different x and y distances, like AutoCAD. Polar snap is used for 3D sketches: distance and angle. Snap to grid is available.

Grid — major and minor grid distances can be specified in Sketch environment only, and then only uniformly (no separate x and y grid distances). The grid is turned on and off through the Tools | Applications Settings dialog box. The range of the grid distance is limited between 1 and 512 "units."

Lineweights — lineweights can be displayed as in AutoCAD, by selecting from a fixed list of widths. Inventor has a second simplified method, where three different lineweights are displayed according to a range. For example, all lineweights less than 0.16" are displayed by a thin line; those over 0.048" by a thick line.

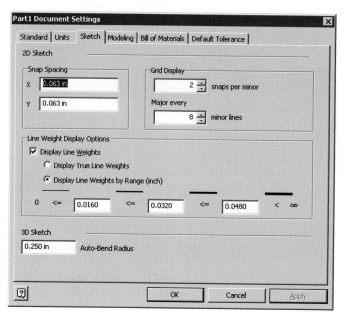

Units — inches or millimeters units are assigned when you choose the template drawing. Through the Document Settings dialog box, however, you can change the units to feet, centimeters, meters, or microns.

Angles are always in decimal format, measured either in degrees or radians. Because Inventor deals with motion, you can specify units for time (seconds or minutes) and mass (lbmass, slug, gram, or kilogram). Units and angles are found in the Units tab.

Materials — just as in AutoCAD, you can specify a default material. The default is named "Default" appropriately enough, and is medium gray in color. Materials are found in the Standard tab, with a long list ranging from ABS Plastic to Wrought Iron.

Unlike AutoCAD, assigning materials also assigns mass properties, such as density and elasticity. These properties are used for analyses, such as finite element analysis. Also unlike AutoCAD, the material "look" is independent of the material properties. The technical editor suggests using this feature to give your bronze Olympic medal a gold finish.

(When you pick a material with a shiny finish, like Gold, the reflection you see is the front door of the Oregon offices where Inventor was developed. This can be customized if you know the trick. The technical editor reports that his shows him water skiing.)

No Layers (kind of), Osnap, Limits, Elevation, Thickness

Inventor has layers, but hides them from you. Only the Drawing environment has a layers droplist for selecting the current layer name. For example, there are predefined layers named "Sketch Geometry," "Centerline," and "Hatch."

In other environments, Inventor places objects on predefined layers automatically through object types. This occurs as you select commands to draw objects, and is controlled by the Style And Standard Editor dialog box.

To access this dialog box, choose **Format** from the menu bar, and then **Style and Standard Editor**. In the dialog box, choose **Layers**. Notice the list of layer names and properties. (See figure below.) You can change the following properties: On (Off), Color, Linetype, Lineweight, Scale by Lineweight, and Plot.

Missing are the Freeze and Lock settings familiar to AutoCAD users.

Unfamiliar to you may be the "Scale by Lineweight" item, which scales the linetype pattern (dots, dashes, and gaps) according to the width of the lineweight.

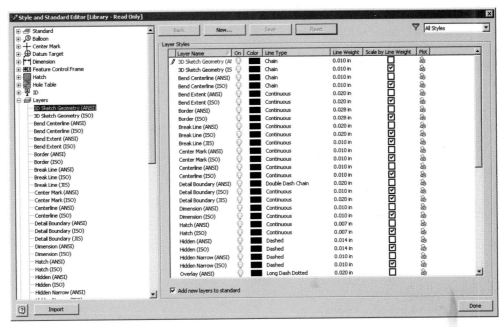

You can add layer names by clicking the **New** button, and remove them by right-clicking and selecting **Delete**.

To determine which entities end up on which layer, Inventor refers to a l-o-n-g master list found under Object Defaults in the same dialog box. This list determines the object style and layer name for every possible object, ranging from 3D Sketch Geometry to Work Points.

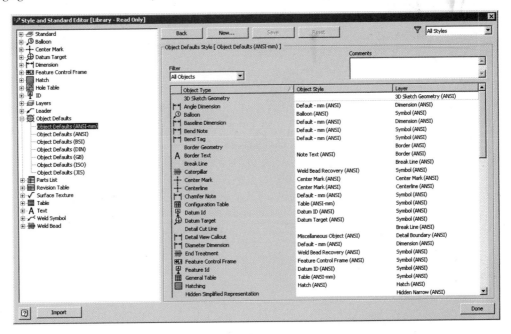

I recommend leaving this list alone. But if you are tempted to play around with it and found you've mucked it up, click the handy **Reset** button to return the list to normal.

Of course, layers are needed for compatibility with AutoCAD. When you import an AutoCAD drawing, Inventor adds the layers to its own.

Other aspects of AutoCAD that Inventor doesn't have (or need):

> **Object Snaps** — instead of object snaps, Inventor uses constraints. Whereas object snaps are used to place geometry in AutoCAD, constraints are used to keep geometry connected together.
>
> When the cursor is close to a geometric feature during drawing commands, Inventor displays a small green dot. Dashed lines show relationships, like AutoCAD's OTrack (object tracking). The icon that looks like an arc in the figure below foreshadows the constraint that will be applied when you complete the line segment.

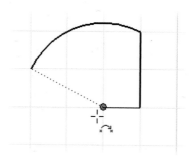

> **TIP** *The following constraints are available in 2D and/or 3D sketch modes:*
>
> - *Perpendicular*
> - *Colinear*
> - *Smooth*
> - *Vertical*
> - *Reference*
> - *Tangent*
> - *Horizontal*
> - *Concentric*
> - *Symmetric*
> - *Pattern*
> - *Coincident*
> - *Parallel*
> - *Equal*
> - *Fix*

> **Otho Mode** — when a line or other object is approximately vertical, Inventor nudges the line to the vertical. Same for the horizontal.
>
> **Limits** — there no need for limits in Inventor, nor in AutoCAD any longer.
>
> **Elevation** and **Thickness** — elevation and thickness are holdovers from AutoCAD's early days; elevation continues to be the substitute for the z-coordinate, while thickness is the clunkiest way to turn 2D objects into 3D.
>
> There is no elevation in Inventor, because it is a true 3D modeler, and so works with x, y, and z coordinates — except in 2D Sketch mode, where everything is truly flat.
>
> Inventor's nearest equivalent to thickness comes into play when extruding 2D objects; the default is 1".

Hatches

Although there is a Fill/Hatch Sketch Region command in Inventor, you (almost) never use it, except maybe to touch up a drawing. That's because hatching is fully automatic in Inventor. Available in the Drawing environment only, hatches are applied to cut-aways automatically, correctly match the assigned material, and are drawn at the right scale.

For example, the bearing made of steel shows the ANSI 32 hatch pattern in section drawn at the correct scale factor. Inventor drew the hatching based on the material assigned to the parts, the type of view (section), and the scale assigned to the view.

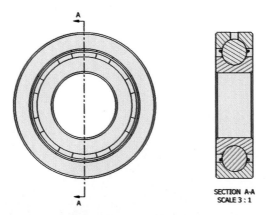

Patterns are hardwired into Inventor, and so you cannot add more. Inventor cannot read AutoCAD's *.pat* hatch pattern definition files. However, it does include a number of patterns that are identical to AutoCAD's.

Inventor's 32 predefined hatches are shown in the dialog box below. (I edited the dialog box so that it shows all hatch patterns at once.)

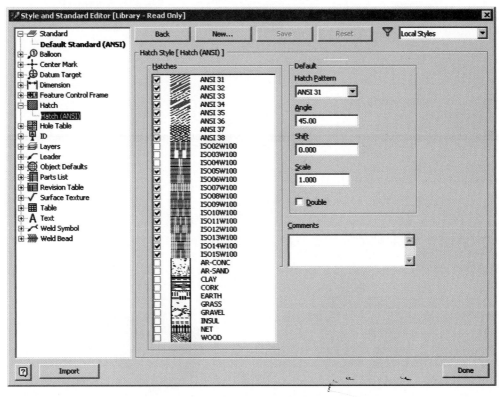

You can mildly customize Inventor's hatch patterns, along the lines of AutoCAD's user-defined patterns: change the angle, double the hatching, and so on.

Visual Styles and Shadows

Inventor has three display modes, which AutoCAD calls visual styles. In the Part and Assembly environments, the three display modes are as follows:

- **Shaded** is the equivalent to AutoCAD's Realistic visual style.

- **Hidden Edge** shows hidden lines in shaded mode, and is equivalent to AutoCAD's Realistic with hidden lines turned on.

- **Wireframe** is equivalent to AutoCAD's 3D Wireframe visual style.

There is no conceptual (Gooch) visual style in Inventor.

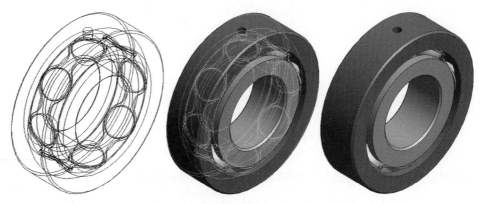

Left: *Wireframe display mode in the Assembly environment.*
Middle: *Hidden edge display.*
Right: *Shaded display.*

You access them from the Standard toolbar:

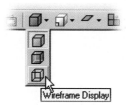

The available display modes vary depending on the environment. Here's what the Drawing environment's display modes look like:

Left: *Hidden line display mode in the Drawing environment.*
Middle: *Hidden line removed.*
Right: *Shaded.*

And, as in AutoCAD, objects can display ground shadows. Better than AutoCAD, however, Inventor has two types of ground shadow: normal and xray (as illustrated below).

As in AutoCAD, you can toggle the view between parallel and perspective projections

I find that Inventor generates visual styles and shadows more easily than does AutoCAD; i.e., there seems to be less strain on the computer, because the effects are generated effortlessly.

How Drawing Commands Are The Same and Different

It is completely possible to replace AutoCAD with Inventor for most 2D drafting tasks. The drawback would be DWG compatibility with other AutoCAD users, because .*dwg* files created in Inventor cannot be fully edited in AutoCAD. (The benefit, retorts the technical editor, is that the designer create assemblies and then generate 2D drawings for AutoCAD users, content in the knowledge that the design cannot be corrupted by them.)

Still, if you just need a 2D drafting tool, just get into Sketch mode and begin drawing and editing. Here's how:

1. Start Inventor.
2. Open a new drawing (*.*dwg*).
3. Click the **Sketch** button.
4. Start drawing!

This creates AutoCAD-compatible drawings, just as if it had come from AutoCAD.

(This talk of not needing AutoCAD, of using Inventor to view .*dwg* files and draft AutoCAD-style 2D drawings, is somewhat silly, because the Inventor package includes a full copy of AutoCAD. The inclusion of Series in the name Inventor Series tells you that AutoCAD is included on the DVD.)

Arcs and Lines

While Inventor's 2D drawing commands are not as comprehensive as AutoCAD's, you may well find them easier to use. Take arcs, for example. Where AutoCAD has eleven ways to draw arcs, Inventor has three:

Inventor	Nearest AutoCAD Equivalent
Three-point arc	Start, second, end
Center-point arc	Center, start, end
Tangent arc	Continued

(Have you ever used all eleven arc methods in AutoCAD? I thought not.) Likewise, there are just three ways to draw circles in Inventor: center-radius, three-point, and tangent-tangent-radius.

Inventor switches automatically between the Line and Arc commands; start a line, select the endpoint, then click and drag in an arc-like movement. Inventor adds an arc tangent to the line's endpoint.

No Polylines

There is no polyline object. Instead, Inventor draws closed objects, such as rectangles and polygons, using lines and arcs that are kept together with constraints. This approach can lead to unexpected behavior, as described later.

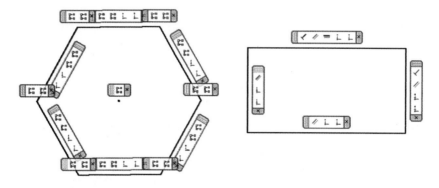

Dimensions and Constraints

Inventor works with two kinds of dimensions: dimensions that *define* parts, and dimensions that *annotate* parts. As an AutoCAD user, you are familiar with the latter: once the drawing is finished, you then add dimensions so that others can read it more easily. As for the former, sketches, features, and so are dimensioned from the get-go.

> **TIP**   *Dimensions are placed in only two of Inventor's environments: Sketch and Drawing. In sketches, they define sizes of objects; in drawings, they annotate objects.*

When dimensioning sketches in the Part environment, you can apply them manually using the **General Dimension** command, or have Inventor do it for you with the **Automatic Dimension** command. I'll talk about general dimensions first, since they are more familiar to you, and far more commonly used.

General Dimensions

In the panel, click **General Dimensions**, and then just select an element, and Inventor knows what kind of dimension to apply:

- Pick a line, and Inventor applies a horizontal or vertical linear dimension. Before placing an aligned dimension, right-click and choose **Aligned** from the shortcut menu.

- Pick a circle, and Inventor applies a diameter dimension; an arc, a radius dimension.

- Pick a vertex, and Inventor applies an angular or linear dimension, depending on your next pick points. Inventor also applies angular dimensions between two lines when you pick the middle third of them.

- Pick the end points of lines, and Inventor applies a linear dimension between them.

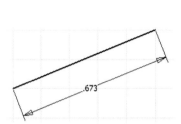
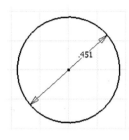
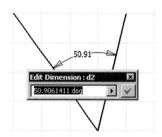

Left: Dimensioning a line.
Center: Dimensioning a circle.
Right: Dimensioning a vertex.

If AutoCAD were only so convenient, especially considering that Mechanical Desktop (which is built on AutoCAD) has had this feature right from its first release. A single command in Inventor rules over all (manually-placed) dimensions. In particular, Inventor knows which dimension to apply based on the object type.

When you try to add a dimension where an identical one already exists, Inventor displays a warning dialog box to let you know the dimension is unnecessary.

After you place the dimension, a small dialog box pops up where you can edit the text. Dimension text can be numbers (as in AutoCAD), variables in equations, or parameter files. When dimensions are determined by equations, the text is prefixed with **fx**. An equation can be as simple as **d2=d1**, where the length of one dimension is tied the length of a second dimension.

Using the General Dimension tool, you can place the following kinds of dimensions in the Sketch, Assembly, and Drawing environments:

- **Linear** dimensions of one object, or between two elements.

- **Aligned** dimensions between two elements.

- **Angular** dimensions between two edges, three points, from reference lines, and of interior or exterior angles.

- **Radial*** and diameter dimensions**.

  <u>Notes</u>:
  *) 3D sketches are limited to linear and angular dimensions; radial dimensions are applied only to bends.
  **) Diameter dimensions are created when centerlines are included in the dimensions.

Automatic Dimensions

Dimensions are closely tied to constraints. Very closely. In fact, Inventor won't even let you add a hole to an object without specifying how far it is located from two edges (x and y distance).

Inventor's **Automatic Dimension** command fully dimensions sketches — fully dimensioned, according to how the sketch needs to be dimensioned for Inventor's purposes. For instance, here is a trapezoid I've dimensioned in the manner to which we have become accustomed: the top and bottom are dimensioned, as are the height and the angles.

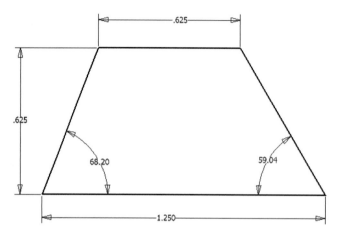

Below is a copy of the same trapezoid that I've allowed Inventor to dimension automatically. It appears that there are too few dimensions. You might think (or cry out loud), "Doesn't it need a dimension here or there and over here? And what about this weirdly placed dimension?" But Inventor placed the dimensions to suit its needs.

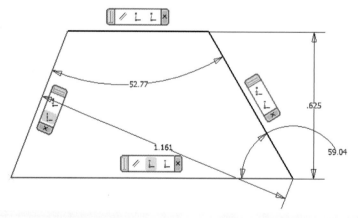

On the other hand, automatic dimensioning is useful as a last resort when you cannot see which dimension is missing, in order to have a fully-constrained sketch.

In addition to dimensions, Inventor also works with constraints, which tend to stay hidden from view, because they tend to clutter drawings. Press **F8** to see them. Constraints are added automatically and invisibly by Inventor as you draw. They give Inventor information about your sketch that obviate the need for dimensions.

In the figure above, constraints show that the four line segments are joined at the vertices. (I've highlighted a pair of them.) Two other constraints show that two line segments are parallel to each other. If you were to add more elements to the sketch, then Inventor would silently determine constraints to them, as well.

To ensure sketches actually work (are the correct size), Inventor determines whether they are over, fully, or under constrained:

> **Over constrained** — the sketch must not be over-constrained (i.e., have too many dimensions), because that would lead to some contradicting each other. When a drawing contains too many dimensions, Inventor asks if you want it to turn surplus ones into *driven* dimensions and indicates this by placing parentheses around the text, like this: **(0.625)**. Driven dimensions cannot be edited until you turn off Driven Dimension on the toolbar.

> **Fully constrained** — the sketch has exactly the correct number of dimensions and constraints.

> **Under constrained** — the sketch has too few dimensions and constraints, and so some elements have sizes that are ambiguous. This is also known as having "degrees of freedom."

Constraints are so important that their status rates a spot on the status bar.

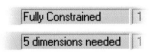

Top: *Fully constrained drawing.*
Above: *Five more dimensions needed to be fully constrained.*

Press **F9** to turn off the display of constraints.

The opposite of constraint is *adaptive*. When elements are adaptive, they can adapt to geometric changes made to other parts in assemblies. For example, you could edit the size of a gearbox and its mounting plate would update accordingly — as would the assembly and the 2D drawings of the gearbox and so on.

Dimension Standards

The look of dimensions is specified by the drawing standard: ANSI, ISO, BSI, JIS, DIN, or GB. As far as I can tell, there are no dimension overrides — other than for how tolerances are displayed.

If you have built up a library of dimensions styles in AutoCAD, then you'll be glad to know that you can import them into Inventor.

Dimension styles are called "Dimension Properties" and sports a multi-tabbed dialog box like the one in AutoCAD. The standards can be modified through the **Tools | Styles And Standard Editor** dialog box.

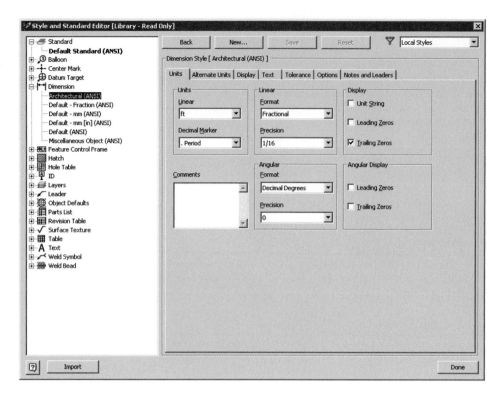

Editing Dimensions

To edit dimension text, double-click any part of the dimension. A small dialog box appears where you change the number. Because dimensions are tied to objects, changing the number resizes the object. That means you cannot replace the number with letters, unless they are related to formulas.

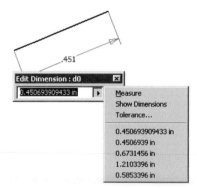

Click the arrow button to select different kinds of measurement:

- **Measure** measures the distance between two elements, between any two points, positions, or element. For example, if you pick a circle, then the circle's diameter becomes the dimension's new measurement.

- **Show Dimensions** toggles the display of dimensions.

- **Tolerance** displays a dialog box for specifying tolerance text.

Differences in Editing Commands

The whole objective of parametric modeling is *never* to edit sketches directly. Instead, you edit them by modifying the value of their attached dimensions or the geometry to constrain. Well, almost never.

There is the occasional instance when you might want to edit sketches in an AutoCAD-like manner. You would tend to use visual editing in Inventor, rather than editing commands.

To visual edit objects, grab one, and then move the cursor. Inventor doesn't have AutoCAD's grips, but it does have points showing the centers of circles and arcs. (Selecting splines shows their control points.) Dragging part of an object logically changes it in the way you would expect.

Editing Lines, Circles, and Arcs

Here are examples for lines circles, and arcs:

Line

Dragging the line moves it
Dragging an endpoint moves the endpoint

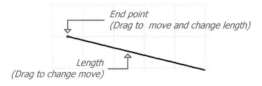

Circle

Dragging the center point moves the circle
Dragging the circumference changes its radius

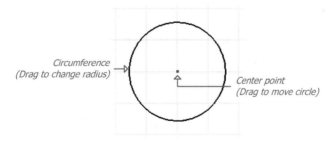

Arc

Dragging the center point moves the arc
Dragging the circumference changes its radius
Dragging an endpoint extends the arc

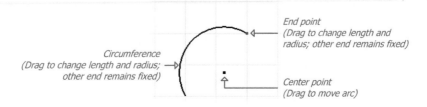

Try it now with a circle:

1. In Sketch mode, use the **Circle** command to draw a circle

2. Move the cursor over the circle. Notice that it turns red, meaning Inventor recognizes you want to edit it.

3. Drag the edge of the circle and the radius changes.

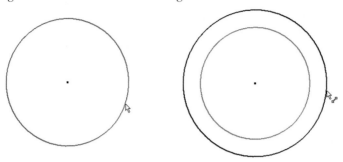

Left: Selecting the circle.
Right: Changing its radius.

As you do this, notice the icon next to the cursor. It indicates the type of editing that's occurring. Above, the double-ended cursor indicates the circle is being stretched.

4. Now move cursor over circle's center point, the black dot. Notice that the center point turns red.

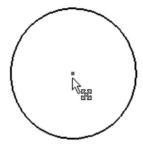

5. Drag it, and the circle moves. The quadruple-arrowed icon means "move."

I discovered exceptions with polygons and rectangles, and found I needed to use different tactics with them. While it seems backwards, it is how Inventor handles polygons:

Polygon

Drag the center to resize it
Drag a vertex or segment to move it

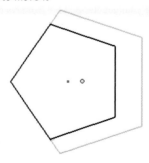

Rectangle

Select the entire rectangle to move it

Drag a segment or vertex to change the size and aspect ratio

> **TIP** *The changes you make to objects are restricted by constraints, and in most cases affect other, attached objects. For instance, when a line is constrained, you may only be able to change its length and not its angle. If a circle is attached to a line, changing its radius may move the line. This "unexpected" behavior depends on the type(s) of constraints applied to objects.*
>
> *To edit an object independently of constraints, you have to delete them: right-click, and then choose* **Delete Constraint***.*

If the object cannot be edited directly with the cursor, then look for additional editing options in the shortcut menu. In this menu, the extra options for editing splines start at Display Curvature.

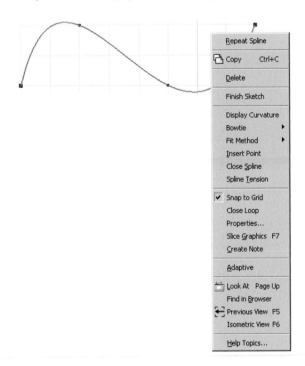

The other way to edit is to choose an editing command from the panel, such as Move or Scale. In this case, Inventor displays a dialog box in which you fill out options. When the arrow button is red, then Inventor needs input from you. Click the button, and then pick a point in the drawing.

For the Move command, there often two red arrows: (1) Select; and (2) Base point. Both are identical to AutoCAD's 'Select objects' and 'Specify base point' options.

The Copy option makes copies.

When you check the Precise Input option, Inventor displays the closest thing you'll ever see to AutoCAD's command line: it lets you enter x, y, and z coordinates numerically.

The **>>** button gives you options for loosening constraints to make movement easier.

Text and Styles

Inventor uses TrueType fonts in drawings, but not SHX fonts. To place text in drawings, enter the **T** shortcut, which displays the Format Text dialog box, one that looks very different from AutoCAD's mtext dialog box.

After text is placed in drawings, simply click it to edit it. (Inventor displays the same dialog box).

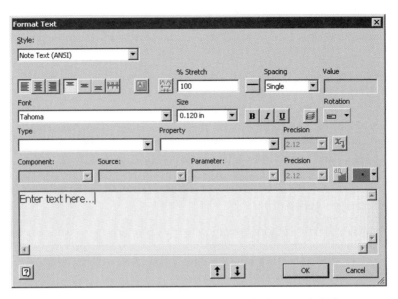

Inventor includes two text styles, the smaller Note Text style and the larger Label Text — one pair for each of DIN, ANSI, JIS, and ISO standard. To create new styles, select **Tools | Styles and Standard Editor**. This opens a monster dialog box that handles styles associated with dozens of items, from balloons to weld beads. Tucked in there are text styles.

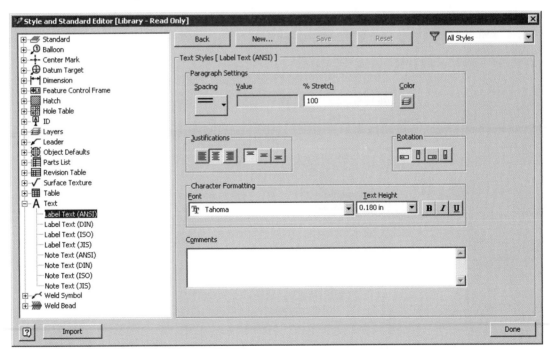

To create a new style, follow these steps:

1. In the **Styles and Standard Editor**, open the **Text** tree.

2. Select an existing style.

3. Click **New**, and then name the new style.

4. Make the changes to the style, and then click **Done**.

The dialog box does not, unfortunately, preview styles, so you can't tell what the font will look like until you use the Text command. To delete a style, right click its name in the tree, and then select **Purge Style** from the shortcut menu.

Spell checking appears unavailable in Inventor, and AutoCAD's field text seems unsupported.

Attributes & BOMs

Only until AutoCAD 2008 did Autodesk properly fix the Bill of Materials (BOM) generator. You use attributes and object properties to generate the BOM as a table in the drawing or in external files.

In 3D, Inventor uses properties in place of attributes. In 2D, sketch symbols and title blocks have AutoCAD-style attributes. And it certainly generates BOMs (from the Assembly environment).

ContentCenter & Design Accelerator

The most important way to be efficient with Inventor is to reuse parts. This is common in mechanical designs, because products tend to use industry-standard parts. They use standard screws and wires. The designs use power cords and rubber feet that tend to be the same. For example, Inventor's Cable & Harness Library includes cables manufactured by Belden, Alpha, and GXL.

In AutoCAD, it is more common to create entire drawings from scratch. You might even have been given assignments in your CAD training classes to draw bolts stylized and with threads. But you would *never* design bolts in Inventor — unless you were a bolt manufacturer; you always work with standard bolts, and then check that the bolts would hold together your assembly with stress analysis.

The Inventor package includes at least 1.5 million parts that are based on standard items from industry. You access them using Content Center (or View Catalog). It seems that it isn't as sophisticated as AutoCAD's DesignCenter, but it doesn't need to be. For example, Inventor has no need to access block definitions, linetype styles, etc.

The Content Center has geometric shapes (like angles and rectangular tubes), pockets and bosses, punches, and slots. You can use the iPart Factory to create thousands of variations on one part and a spreadsheet.

The second source is Design Accelerator, which generates parts from your specifications. If you need bolts, you input the following parameters: diameter, number of bolts and material. Inventor figures out the needed length. Then you check if the bolts are strong enough for the job.

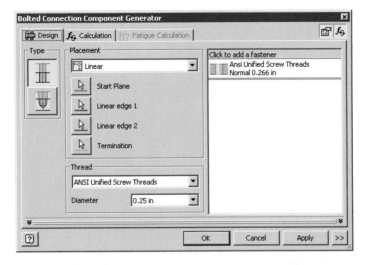

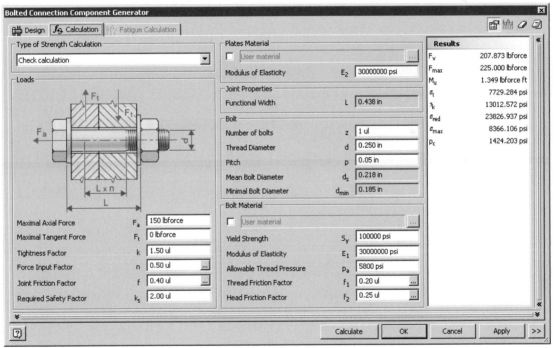

Another example: using the Design Accelerator you can specify the power, speed, and drive ratio. Inventor calculates the required size and type of belts and pulleys, rounding things off to suit industry-standard stock components. It then inserts the components into the assembly. When you edit a specification, such as power or speed, then everything is updates automatically.

Design Accelerator is available for generating bearings, spur gears, bolted connections, clevis pins, chains, v-belts, keys, solder joints, beams and columns, shafts, and more. Select **Design Accelerator** from the Panel's title bar in the Assembly environment.

For other disciplines, Inventor also includes calculators for generating tubes, wiring harnesses, electrical connectors, and more.

When you get beyond a few dozens of drawings, you need ways to keep track of them. Autodesk provides software and services for doing just that.

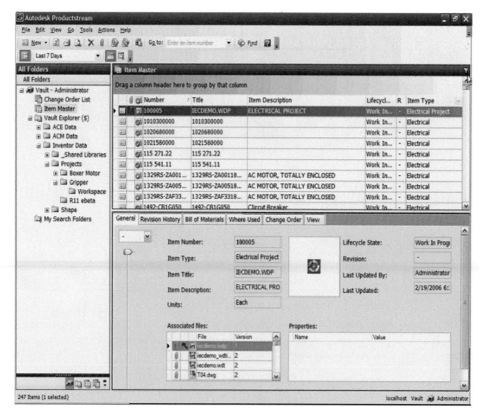

Design Assistant — Inventor's file tracking service. Because all of Inventor's files are interconnected, you cannot arbitrarily rename or move files without breaking the links among them.

Streamline — Autodesk's project management service. Whereas Buzzsaw is for construction projects, Streamline is for products designed in Inventor. Both are central repositories for drawings and specifications, as well as tracking the progress of projects. Files are displayed in DWF format. You can learn more about it from www.autodesk.com/streamline; the service is not free.

Vault — similar to Streamline, but runs inside your office. It tracks drawings, does revision control, and ensures that components are reusable. The software is included free with Inventor. Learn more from www.autodesk.com/vault; alternatively, from Inventor's **Help** menu, select **Vault Help**.

ProductStream — grown-up version of Vault. It adds automated releases, bills of materials, and engineering change order management. You can learn more about it from www.autodesk.com/productstream; the software is not free.

Plotting Similarities and Differences

The similarity between AutoCAD and Inventor for plotting drawings is that they both do it. The difference is that Inventor dispenses with AutoCAD's (some would say horrifically complex) plotting options, and instead uses a bare bones approach.

Print Setup

The Print Setup (a.k.a. plot options) dialog box is so simple I have to wonder if it is borrowed from Notepad. To see it, choose **Print Setup** from the **File** menu:

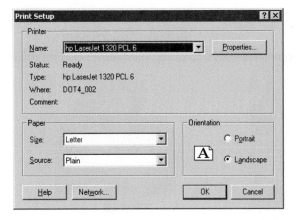

Inventor works with System printers and plotters, those that have drivers provided by Microsoft or the vendor. There are no Inventor-specific plotter drivers, as with AutoCAD.

A different, more useful print setup dialog box is displayed when you instead choose the **Print** command.

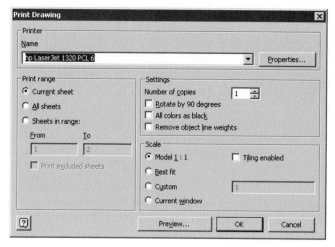

Click **OK** to print the drawing(s).

Plot Stamps

There is no PlotStamp command in Inventor, but the facility exists. Set up the title block to display properties extracted from the file properties. This is like using field text in AutoCAD.

Print Preview

Print preview has a somewhat different user interface from AutoCAD. There is no right click menu, and you can't press Esc to exit. Instead, make use of the row of buttons along the top.

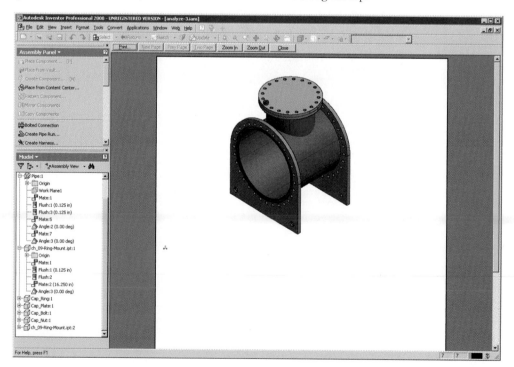

When you click **Print**, you see the Print Drawing dialog box mentioned earlier.

> **TIP** To save a raster image of 3D models, use the **File | Save Copy As**
> command. In the **Files of Type** droplist, choose a raster format, such as TIFF or
> JPEG.

Publishing 3D Models

The **Publish** command outputs the drawing(s) as *.dwf*, just as in AutoCAD. Inventor publishes assemblies and i-assemblies, parts and i-parts, drawings, weldments, sheet metal parts, and presentations in DWF format.

From the **File** menu, select **Publish**. The dialog box looks quite different, however, and has fewer options. That's because all DWFs are 3D, and because layers are not as important in Inventor as in AutoCAD. This depends on whether you are publishing a 3D model or a 2D drawing.

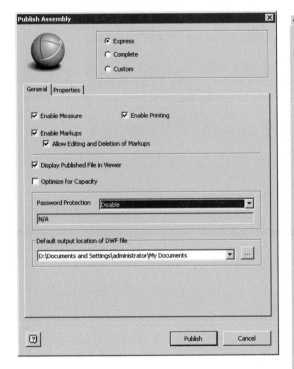

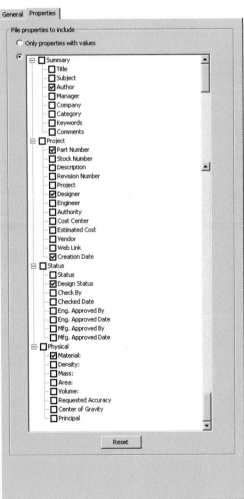

The Custom preset adds a tab to this dialog box that contains options specific to the type of drawing. Here's the extra tab for assemblies:

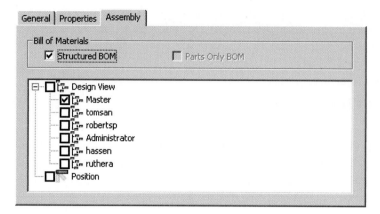

All of the added options are as follows:

Drawings — choose the current sheet only or all sheets.

Assemblies and **Weldments** — choose the design views, positional representations, and bills of material.

i-Assemblies — choose the factory components, 3D models, design views, positional representations, and bills of materials.

i-Parts — choose the factory components and 3D models.

Sheet metal — choose a sheet metal style, part, a flat pattern to be published.

Presentations — choose views with animations and text info.

Autodesk notes that DWF viewers have these limitations when displaying Inventor models:

- Sketch and work geometry is not published.
- Decals and multiple textures are not displayed.
- Threads are concentric, not spiral.
- Default lighting is used in the DWF viewer.

Summary

This completes the comparison between AutoCAD and Inventor. This comparison, however, is by no means complete, for both programs carry many more features that are the result of a decade or more of programming and user requests.

In the next chapter, you learn how to sketch and model with Inventor — side by side with AutoCAD. The tutorial shows the unfamiliar steps in Inventor, along with the familiar equivalents in AutoCAD.

Part II

2D-to-3D Tutorials

Chapter 4

Sketch to Part Tutorial

In this tutorial, you create a 3D *part* from 2D drawings (*sketches*). The part is a clevis, as illustrated on the following page.

Creating parts in Inventor typically follows the same path each time. Learn from this tutorial, and you can tackle most any other kind of pars. (Recall that a part is of a single 3D object, whether it looks simple or complex.)

These are the steps to creating nearly every kind of part with Inventor:

1. Create a rough 2D sketch of the part's base profile.
2. Add constraints until sketch is fully constrained.
3. Place dimensions manually, and then edit them to precise values.
4. Extrude the part to 3D; revolve, sweep, loft, and other features are also available.
5. Repeat the above steps for other profiles, using them to form the part into the desired shape.

IN THIS CHAPTER

- Sketching in 2D.
- Using constraints and dimensions.
- Extruding sketches to 3D features.
- Knowing common keyboard shortcuts.
- Inserting holes and threaded holes.
- Attaching work planes.
- Adding fillets.
- Assigning material properties.

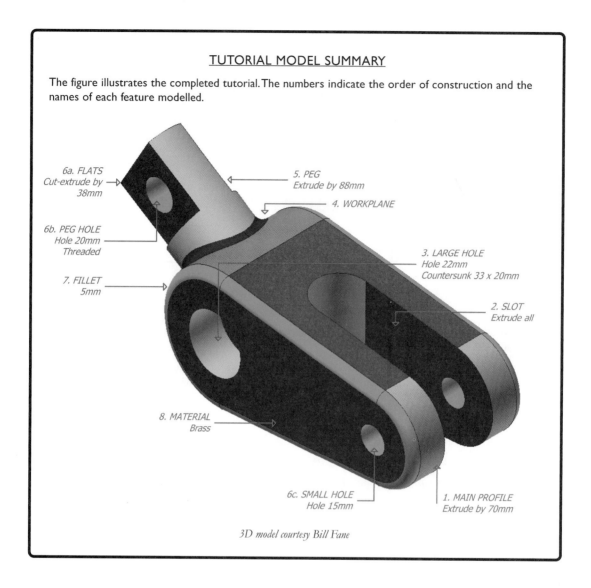

TUTORIAL MODEL SUMMARY

The figure illustrates the completed tutorial. The numbers indicate the order of construction and the names of each feature modelled.

6a. FLATS Cut-extrude by 38mm

5. PEG Extrude by 88mm

4. WORKPLANE

6b. PEG HOLE Hole 20mm Threaded

3. LARGE HOLE Hole 22mm Countersunk 33 x 20mm

7. FILLET 5mm

2. SLOT Extrude all

8. MATERIAL Brass

6c. SMALL HOLE Hole 15mm

1. MAIN PROFILE Extrude by 70mm

3D model courtesy Bill Fane

Stage 1: Main Profile

In the first stage of this tutorial, you sketch the base profile, and then extrude it to a part. See "1. Main Profile" in the figure above.

1. Start Inventor by double-clicking the icon on your computer's desktop.

Autodesk Inventor
Professional 2008

2. (If necessary, click the **New File** button in the Quick Launch area.)

   a. In the New File dialog box, click the **Metric** tab.

   b. Choose "Standard(mm).ipt" to open a part template in metric units.

   c. Click **OK**.

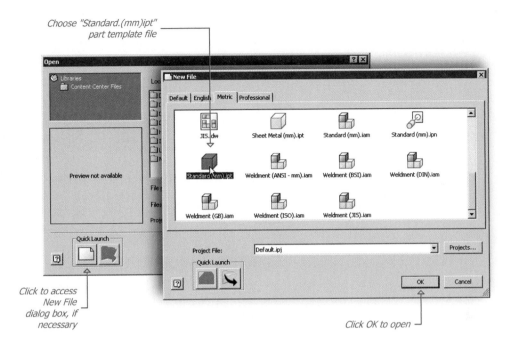

Choose "Standard.(mm)ipt" part template file

Click to access New File dialog box, if necessary

Click OK to open

Notice that Inventor opens a new part file (*Part1.ipt*), and then starts in sketch mode.

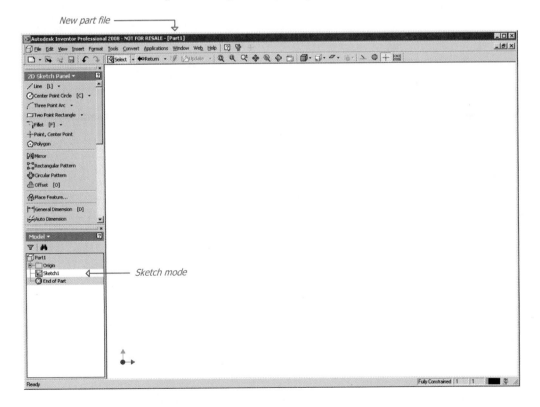

New part file

Sketch mode

In the first stage of the tutorial, you draw the main *profile*, illustrated in blue below. To create the profile, you draw the *sketch*, shown in red. The profile is made of two circles and two lines. Sketch them, as follows:

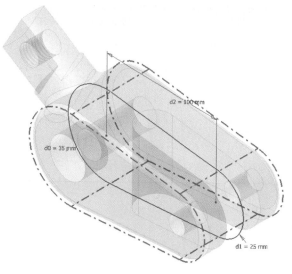

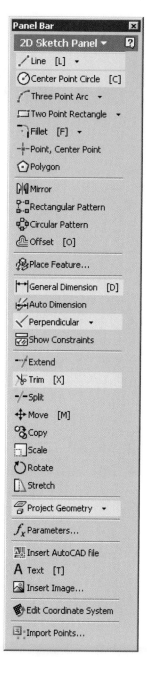

3. To draw the first circle, click **Center Point Circle** in the 2D Sketch panel. (The commands you'll be using are found on the panel shown at right.)

   As you move the cursor in the drawing area, notice the yellow dot that dogs the cursor. It shows the nearest snap point, which is just like snap spacing in AutoCAD.

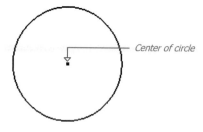

Crosshair cursor (+) ⟶ — *Snap (yellow dot)*

4. Notice the AutoCAD-like prompt on the status bar:

   Select center of circle *(Pick a point.)*

   Pick a point anywhere in the left half of the drawing area.

5. At the next prompt, make the circle any diameter; later, you will use dimensions to size it correctly.

   Select point on circle *(Pick a point.)*

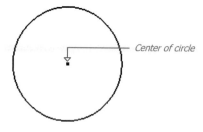

— *Center of circle*

Notice the black dot that indicates the circle's center. You will make use of it in drawing a second circle.

> **TIP** *Just as in AutoCAD, commands repeat themselves until you pick another command or press **Esc**.*

6. To draw the second circle in-line with the first, follow these steps:

   a. Move the cursor to the center of the circle. Notice the green dot and the arc-like icon:
      Green dot — Inventor found a geometric feature (the circle's center).
      Arc-like icon — Inventor will apply the *coincident constraint*, kind of like an object snap with glue. (You learn more about constraints later in this tutorial.)

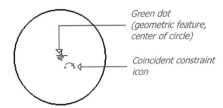

*Green dot
(geometric feature,
center of circle)*

*Coincident constraint
icon*

   b. Without clicking a mouse button, move the cursor horizontally.

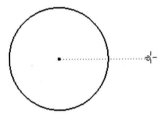

   Notice the dashed line that appears; this is just like AutoCAD's object tracking. (You constraint it horizontally later.)

   c. Click to start the second circle, and then click a second time to give the circle its diameter. Make the second circle about half the size of the first; if the circles are too similar, Inventor may be tempted to automatically place horizontal or parallel constraints — leading to extra work later.

   You now have two circles in the drawing.

7. Add the two lines: from the 2D Sketch panel, choose **Line**.

   a. Draw a line from the top of one circle to the other. To ensure that the line is attached to (coincident with) the circle, touch the circle with the cursor. When you click the left mouse button, be sure that the yellow dot snaps to the circle.

*Tangency
constraint
icon*

   b. Draw the second line between the bottoms of the circles.

If you are fortunate, Inventor creates *tangency constraints*; if not, you'll add them in a later step. Tangency is shown by the circle icon, as illustrated above.

(Good practice is to apply geometric constraints before applying dimensions, note the technical editor.)

8. The lines need to be tangent to each circle so that they stretch correctly later when you size them with dimensions. To ensure the lines are tangent to the circles, take these steps:

    a. Click the **Perpendicular** droplist in the 2D Sketch panel, and then choose **Tangent**. Notice the tangency icon near the cursor.

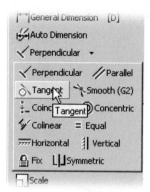

    (This droplist shows the name of the last constraint used, so a name other than "Perpendicular" might show up. In any case, the constraint droplist always appears between Auto Dimension and Show Constraints.)

    b. Pick a circle.

        Select first curve *(Pick a circle.)*

        Notice that it turns red to indicate it is selected. This is the equivalent of highlighting selections in AutoCAD.

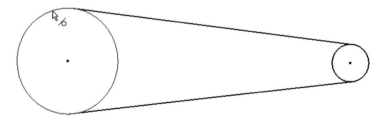

        TIP *Notice that the status bar prompts you to select a curve. "Curve" is Inventor's generic term for 2D primitives, like lines, arcs, circles, and splines.*

    c. And then pick the nearby line.

        Select second curve *(Pick a line.)*

        If Inventor complains, "Sketch constraint already exists," then click **Cancel** to dismiss that dialog box.

    d. Repeat until you have touched all four tangencies. (See the tip below on how to verify the existance of tangency.)

CONSTRAINT ICON SUMMARY

Instead of object snaps, Inventor uses *constraints* to snap to geometry and keep sketched objects together. Before a constraint is created, Inventor displays its icon near the cursor. Unfortunately, Inventor doesn't display tooltips that explain the meaning of the icon, so here is a summary chart.

The first column of icons (at the left) are the constraint icons shown by panels. The second column (to the right) are the icons displayed at the cursor.

Perpendicular: lines, ellipse axes, spline handles, text edges, and images.

Tangent: curves and ends of splines.

Coincident: two points, or a point and a curve.

Colinear: lines, ellipse axes, spline handles, text edges, and images.

Horizontal (*used in 2D sketches only*): lines, pairs of points including arc and circle centers, ellipse axes, spline handles, text edges, and images.

Parallel: lines, ellipse axes, spline handles, text edges, and images.

Smooth (*G2*): between splines and lines, arcs, and other splines.

Concentric (*used in 2D sketches only*): arcs, circles, or ellipses.

Equal (*used in 2D sketches only*): arcs, circles, and lines.

Vertical (*used in 2D sketches only*): lines, pairs of points including arc and circle centers, ellipse axes, spline handles, next edges, and images.

Symmetric (*used in 2D sketches only*): lines or curves.

Fix: points and curves.

Reference: projected geometry.

Pattern (*used in 2D sketched only*): available after rectangular and circular patterns are created.

e. Press **Esc** to exit tangent-constraint mode.

If you wish, grab one circle and drag it around. Notice that the lines stretch and move to remain tangent and attached to the circle.

> **TIP** To see the constraints, press **F8**. Move the cursor over a constraint icon to see the connections it enforces.
>
>   In the figure below, the cursor is on a tangency icon; the two curves tangent to each other are highlighted in red.

> To remove a constraint, select it, and then press **Del**.
>
> Press **F9** to turn off the display of constraints.

9. Trim the circles to create the outline of the main profile:

a. Choose **Trim**.

Select portions of curves to trim *(Pick inner portions of circles.)*

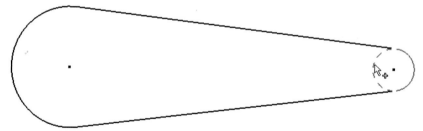

b. Pick the inside parts of both circles to erase them.

Notice that Inventor doesn't ask you for cutting edges; they are explicit, like replying Ctrl+A (select all) at AutoCAD's Trim command's "Select objects or <select all>:" prompt.

> **TIP** In Inventor, you do not need to specify linear or radial dimensions, or use object snaps with dimensions — as you are forced to in AutoCAD. Inventor knows that the dimension of arcs should be radial, of circles should be diametral, between two points should be linear, and so on.

10. The two centers need to aligned. Choose **Horizontal** from the flyout of constraints, and then pick the two center points. Notice that the centers snap to the horizontal.

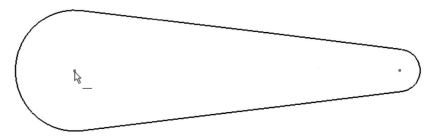

11. Apply dimensions to size it the sketch, as follows:

    a. Click **General Dimensions**.

    b. Choose the arc on the left. Notice the circular icon; this indicates that Inventor has

       detected the object is an arc and will apply a radial dimension. (When dimensioning

       circles, Inventor applies diameter dimensions.)

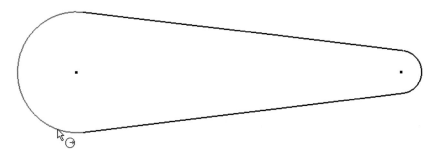

    c. Drag the dimension line away from the arc, and then click.

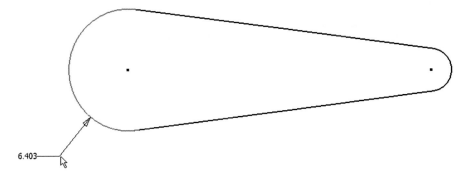

    d. In the dialog box, enter **35**, and then click the ☑ green checkmark button.

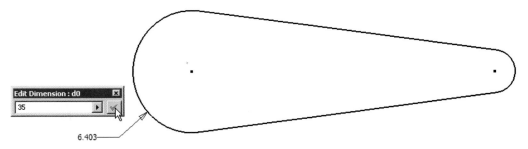

Notice that the arc resizes itself, and that the lines' ends move with the resized arc. You are seeing Inventor's constraints at work, ensuring that the lines and arcs always stay connected to each other.

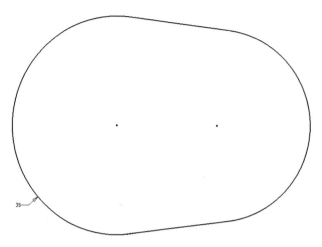

> **TIP** *To see the entire drawing, press **Home**. This is equivalent to AutoCAD's Zoom All command.*
>
> *To give some room around the line, roll the scrollwheel of your mouse. Roll it one way to zoom in, the other way to zoom out — opposite to AutoCAD. (To match AutoCAD, use the Tools | Application Options | Display dialog box's **Reverse Zoom Direction** option.)*

12. Repeat for other circle, but use **25**.

    Press **Home**, if necessary, to see the entire drawing.

13. The distance between the centers of the circles is 100mm. When using the General Dimension command, follow these steps to dimension between two points:

    a. Pick the dot in the center of one circle.

    b. Pick the dot in the other.

    c. Enter **100** in the dialog box, and then click the checkmark button.

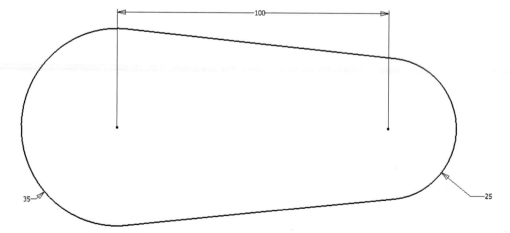

14. Looking at the status bar, you will see Inventor complaining that two dimensions are needed. (If it states that three or more dimensions are needed, then you may have missed setting some lines and circles tangent to each other.)

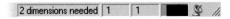

In order to be fully defined, the sketch needs two more dimensions. But what might they be? Inventor wants to know where this sketch is located in relation to the origin, 0,0.

You could apply a couple of linear dimensions between 0,0 and a point on the sketch. But the better approach recommended by the technical editor is to constrain the sketch to the origin. This is just like basing drawings in AutoCAD around the origin to take advantage of the inherent symmetry. Project one arcs's centerpoint to the origin, as follows:

a. From the 2D Sketch panel, choose **Project Geometry**.

b. In the Model panel, expand **Origin** (by clicking the **+** next to it), and then choose **Center | Point**.
Notice that a white dot appears. It indicates the origin, a point to which the sketch can be constrained. (After you exit from the Project Geometry command, the dot turns brown.)

c. From the constraints flyout, choose **Coincident**.

d. Pick the origin point, and then pick an arc's center point.

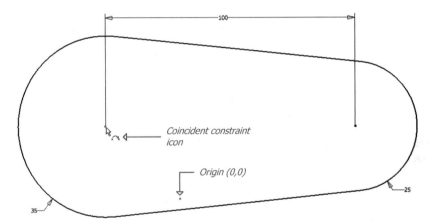

Notice the sketch jumps to relocate itself at the origin, and that the status bar reports,
"Fully Constrained."

15. The sketch for the main profile is now complete, and it is time to extrude it.

To exit 2D sketch mode, click **Return** on the Standard toolbar.

Notice that the panel changes from 2D Sketch to Part Features.

16. Save your work as *Clevis.ipt*, an Inventor part file.

a. From the **File** menu, choose **Save As**.

b. In the Save As dialog box, enter "clevis" for the file name.

c. Click **Save**.

Extruding the 2D Sketch to a 3D Feature

With the 2D sketch done, you can now extrude it into a 3D feature:

1. To see the sketch in 3D, press **F6** , which swivels the view to isometric.

   (Alternatively, right-click and then choose **Isometric View** from the shortcut menu.)

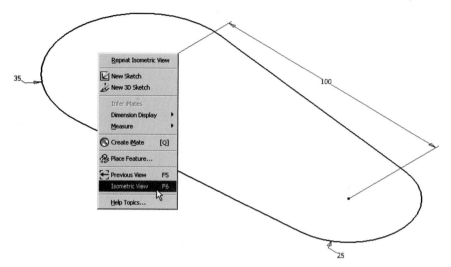

2. From the Part Features panel, choose **Extrude**. Notice the Extrude dialog box.

   The sketch is immediately selected, because it is the only one. Notice that it previews the resultant feature at the default 10mm distance.

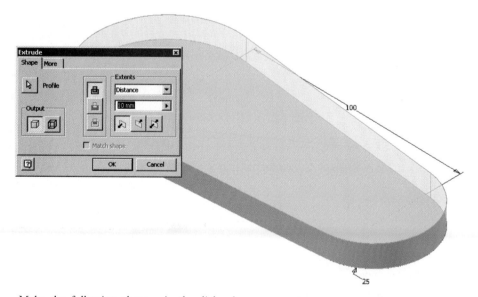

   Make the following changes in the dialog box:

   • Click the [icon] two-sided button; notice that the extrusion now extends in both directions from the profile.

   • Change the distance to **70**; notice that the extrusion thickens. The 70mm is the total extrusion thickness, 35mm in each direction.

- Click **OK**.

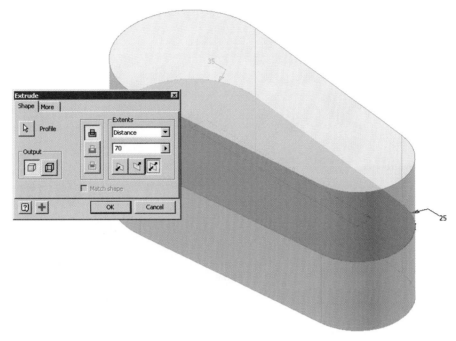

3. Extruding the sketch creates a 3D solid. Its gray color is the default material for Inventor.

   In the Model browser, notice that Sketch1 is now a subset of Extrusion1. When you move the cursor over **Sketch1**, Inventor displays the original sketch in red.

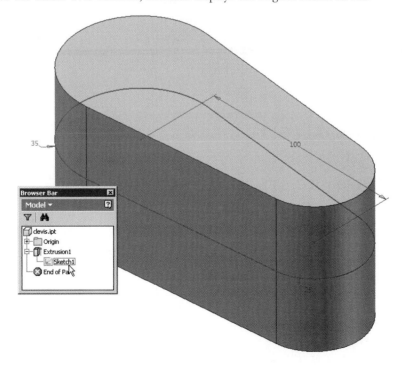

When you move the cursor over **Extrusion1**, Inventor displays the extrusion in red.

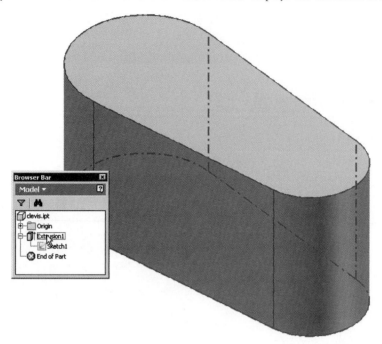

4. Slowly click **Extrusion1** twice in Model Browser, and then rename it to "Main Profile." (Don't double-click! Otherwise, you end up editing the feature.) This helps identify the features that make up the part.

5. Save your work by clicking the diskette icon on the Standard toolbar.

EXTRUSION OPTIONS SUMMARY

The Extrude command's dialog box has these options:

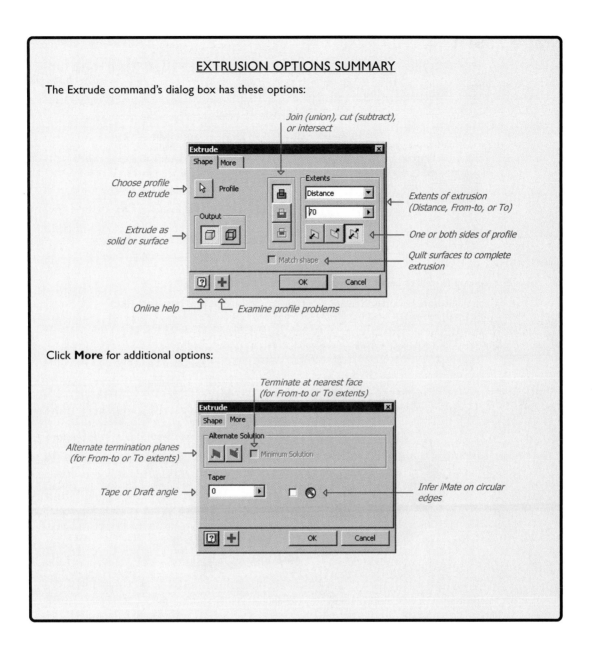

Join (union), cut (subtract), or intersect

Choose profile to extrude →

Extrude as solid or surface →

Extents of extrusion (Distance, From-to, or To)

One or both sides of profile

Quilt surfaces to complete extrusion

Online help — ⌐ ⌐— Examine profile problems

Click **More** for additional options:

Terminate at nearest face (for From-to or To extents)

Alternate termination planes (for From-to or To extents) →

Tape or Draft angle →

Infer iMate on circular edges

Stage 2: Slot

The next stage is to cut the slot from the Main Profile. This too is created from a sketch, which is then extruded and subtracted from the Main Profile.

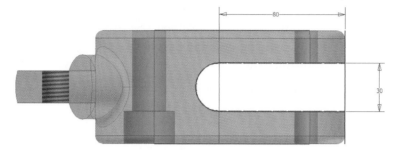

Here's how:

1. First, rotate the view to better position the Main Profile.

    a. Click the **Rotate** button on the Standard toolbar. This command is equivalent to AutoCAD's 3dOrbit command.

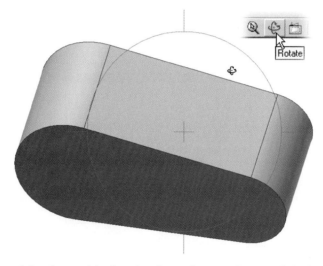

    b. Drag around the view, noticing how it twists and rotates. Inventor is in Free Rotate mode.

    c. Right-click, and then choose **Common View** from the shortcut menu. Notice the 3D view block. The green arrows represent the eight common engineering views: top, left, front, isometric, and so on.

d. Click an arrow, and Inventor rotates the view to match it. The view looks down the selected arrow. This is how you get precise 3D viewpoints in Inventor.

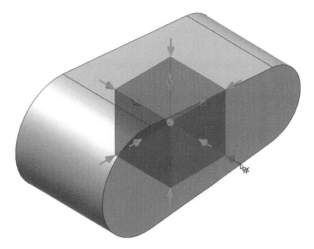

TIP *Press spacebar to switch between Free Rotate and Common View modes.*

e. You can now assign this viewpoint as the default for F6 (Isometric View). Right-click, and then choose **Redefine Isometric** from the shortcut menu.

From now on (and until you redefine it again), pressing **F6** brings up this viewpoint.

2. You can sketch only on flat surfaces. There are no suitable flat surfaces on the part, but you don't need to create a flat surface. You can use a coordinate plane for the sketch plane, which is centered on the origin. Because you positioned the Main Profile on the origin, the coordinate planes intersect the 3D solid.

To help you better see inside the 3D solid, change the display to **Wireframe**.

3. In the Model panel, expand **Origin** (if necessary), and then move the cursor over the three planes: YZ, XZ, and XY.

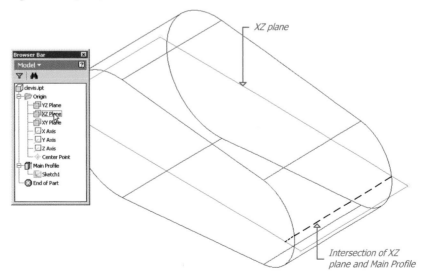

As you do, notice that the planes are outlined in red.

4. Right-click **XZ Plane**, and then choose **New Sketch** from the shortcut menu. (Alternatively, choose **XZ Plane**, and then click **Sketch** on the Standard toolbar.)

Notice that Inventor creates Sketch2 in the Model panel, and goes into 2D Sketch mode.

5. To help create the sketch of the slot, you will now project the intersection of the workplace with the Main Profile onto the sketch plane. This is shown by the dashed line in the figure above.

a. In the 2D Sketch panel, choose **Project Geometry**.

b. The projected edge is unfortunately invisible, so move the cursor around in the general area until a bright red line appears, like the one illustrated below.

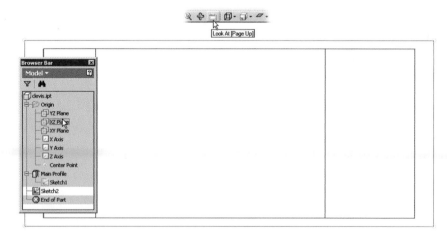

c. Click the red line. Notice that it turns brown. It is now a line that can be used for creating geometry, dimensioning, and constraining. It becomes part of the sketch for the extrusion.

d. Press **Esc** to exit projection mode.

6. Rotate the view so that you can sketch on the XZ plane. The easy way to do this is to look at the plane:

a. In the Standard toolbar, click the **Look At** button.

b. In the Model panel, choose **XZ Plane.** Notice that the view rotates around so that the plane faces you, flat.

(If necessary, press **Home** to make the model fit the screen.)

7. Draw the sketch for the slot, as follows:

   a. Choose **Line**.

   b. Start the line at the newly-created projection line (1).

   c. End the line so that it is perpendicular to the projected line (2). You'll know it is perpendicular by the icon that appears near the cursor.

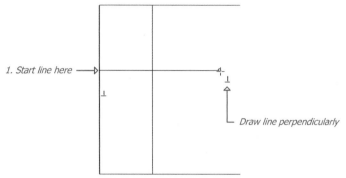

Again, it does not matter where you begin the line, or how long you draw it. These details are handled later by dimensions.

8. Continue with the arc. There is no need to switch to the Arc command. Not at all. Notice the prompt on the status bar:

   Select start of line; drag off endpoint for tangent arc

   a. Holding down the mouse button, drag the cursor. As you do, notice that Inventor drags an arc tangent to the line.

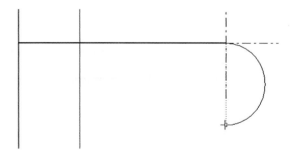

   b. Drag the arc 180 degrees, and then click.

   c. Move the cursor in a straight line, and Inventor switches back to line mode.

   d. End the line back at the projection.

      Press **Esc** to exit the Line command.

You've drawn the sketch of the slot. Now it needs to be positioned with constraints, and then dimensioned.

9. To center the sketch on the plane, you can make use of the arc's centerpoint, like this:

   a. From the constraints flyout, choose **Horizontal**.

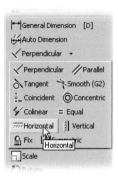

   b. On the status bar, Inventor prompts you:

      Select line, ellipse axis, or first point *(Pick the arc's center point.)*

      Pick the center point of the arc.

   c. At the next prompt:

      Select second point *(Pick line's midpoint.)*

      Move the cursor along the middle of projected line until a green dot appears. This indicates that Inventor has found the midpoint of the line.

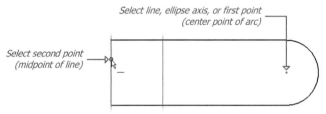

   d. Click at the midpoint. Notice that the entire sketch jumps into place, centered on the sketch plane.

10. With the sketch correctly positioned by the horizontal constraint, you can now dimension it. Choose the **General Dimension** command, and then specify a length of **85**, as illustrated below.

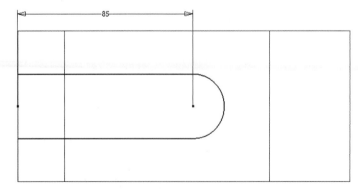

   Notice that the sketch gets longer to accommodate the dimension's demand.

11. In a similar manner, dimension the width of the slot, **30**mm.

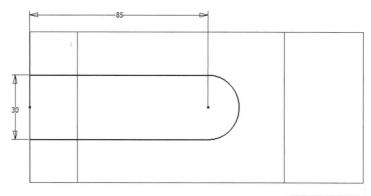

A quick glance at the status bar shows you that the sketch is fully constrained. No more dimensions are needed — or allowed. If you were to add another dimension, Inventor would complain, "Accepting this dimension would over-constrain the sketch."

12. Click **Return** to exit sketch mode.

Press **F6** to see the model isometrically. (As the model rotates, notice that the dimension text keeps facing you.)

13. Choose **Extrude** from the Part Features panel.

This time there is a red arrow next to Profile. This is a mild warning from Inventor, in that it does not know which profile you want extruded.

Move the cursor over sketch 2. Notice that it turns reddish. Click to select it.

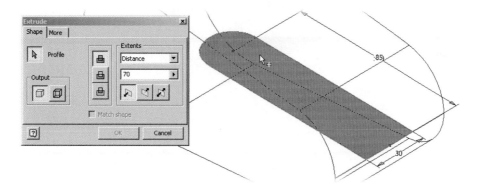

14. Make the following changes to the dialog box:

   • Change **Extents** to "All." This ensures the slot exists, no matter how thick the rest of the clevis may become in later editing operations.

   • Click **Both Ways** to make the sketch extrude up and down.

   • Click **Cut** to subtract the slot from the Main Profile.

Notice that Inventor's Extrude command combines both extrusion and Boolean operations, subtract (cut) in this case. AutoCAD would require two commands: Extrude, followed by Subtract.

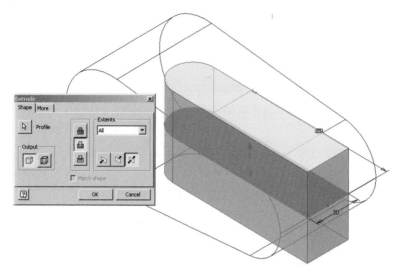

15. Click **OK**, and the slot material is removed. To see the slot better, change the display to **Shaded**.

16. Save your work.

COMMON KEYBOARD SHORTCUTS

Shortcut	Command	Shortcut	Meaning
A	Arc (center point)	L	Line
C	Circle (center Point)	M	Move
D	Draft face	O	Offset
E	Extrude	R	Revolve
F	Fillet	T	Text
H	Hatch/fill sketch region	X	Trim
	Hole (part features mode)	Z	Zoom window

Stage 3: Large Hole

The clevis has a counterbored hole, which you add now. (By adding it at this stage, you make things easier later when it comes time to rewrite history.)

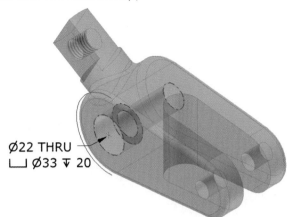

The hole's specification has the following meaning: 22mm diameter all the way through; countersunk 33mm diameter, 20mm deep.

In AutoCAD, you would create countersunk holes by subtracting two cylinders from the solid. Inventor uses the Hole dialog box that generates just about any kind of hole you can think of. Here's how it works:

1. From the Part Features panel, choose **Hole**. (Alternatively, type **H** on the keyboard. Like AutoCAD, Inventor has a number of keyboard shortcuts for commands, as listed on the previous page. Unlike AutoCAD, the command is executed as soon as you press the letter on the keyboard — there is no need to press Enter!)

   Notice the Hole dialog box.

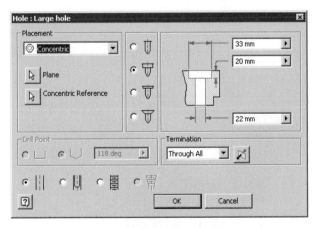

2. Make the following changes, as illustrated above:

   - **Placement** = Concentric (see figure next page)
   - **Bore** = Counterbore
   - **Counterbore width** = 33
   - **Counterbore depth** = 20

- **Hole width** = 22
- **Termination** = Through All
- **Hole type** = Simple

3. With the parameters set up, you specify the location of the hole. The Concentric placement needs to know two things: (1) the face through which the hole is bored, and (2) what the hole will be concentric to. In this case, the center of the hole is positioned by:

- **Plane** = pick the face indicated by "Plane."
- **Concentric Reference** = pick the arc shown in red.

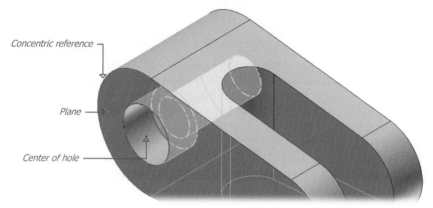

4. Click **OK** and the hole is complete.

Stage 4: Work Plane

To add the peg at an angle, you'll need an angled work plane on which to sketch. Work planes are needed when there is no suitable flat surface on which to sketch.

Work planes attach themselves according to what you pick, such as three points, an axis, or an object. You can see that work planes are somewhat like setting a UCS in AutoCAD. The work plane will be attached to a work line centered on the larger arc of the base feature, and tilted at 45 degrees, as shown below.

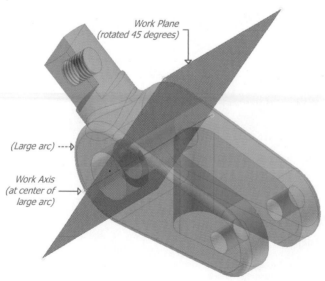

1. From the Part Feature panel, choose **Work Axis**.

   At the prompt:

   Define work axis by highlighting and selecting geometry

   Pick the outer arc (highlighted in red, below). Notice the gold line that appears; it is the work axis.

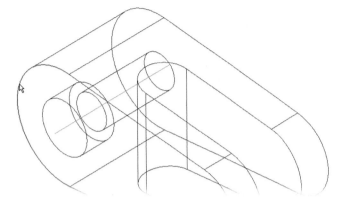

2. From the Part Feature panel, choose **Work Plane**.

   At the prompt:

   Define work plane by highlighting and selecting geometry

   Pick two objects:

   - In the drawing, pick the work axis (the gold line).
   - And in the Model panel, pick **XZ Plane** (under Origin).

   Notice that the work plane appears at 90 degrees, and that an Angle dialog box appears.

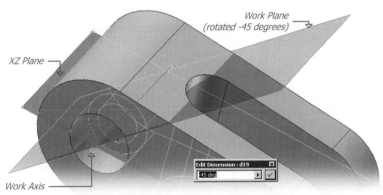

3. Enter **-45** degrees to tilt the work plane, and then click the green checkmark.

   The work plane is in place. Notice that it rates an entry in the Model browser, which means you can edit it like other objects.

4. Save your work.

Stage 5: Peg

The peg is an extruded circle. The circle is sketched on the work plane.

1. In the Standard toolbar, click **Sketch**.

2. Select the work plane. Notice that the Model browser adds Sketch3 to its list.

    If necessary, use **Rotate** to change the view so that you can see the new sketch plane clearly.

3. Draw the circle on the work plane, centered on the drawing's projected center point:

    a. In the 2D Sketch panel, choose **Project Geometry**.

    b. In the Model panel, choose **Center Point** (under Origin).

    c. Back in the 2D Sketch panel, choose **Center Point Circle**.

    d. At the "Select center for circle" prompt, pick the center point. (It shows as a green dot when the cursor gets near to it.)

    e. At the "Select point on circle" prompt, pick anywhere to give the circle an arbitrary diameter.

    f. Press **Esc** to exit the Circle command.

4. Use the **General Dimension** command to set its diameter at **40**mm.

5. Use the **Extrude** command to extrude the circle **88**mm.

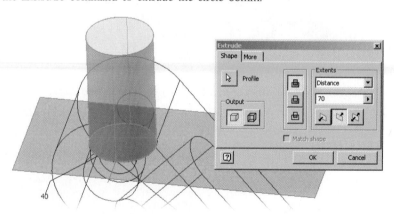

6. Click **OK** to exit the Extrude dialog box.

> **TIP** *You can make the work plane invisible, as follows: In the Model browser, right click Work Plane1, and then choose* **Visibility***. Although it is now invisible, the work plane still exists, and its visibility can be turned on again.*

7. Rotating the 3D model, notice that the peg hangs down into the hole.

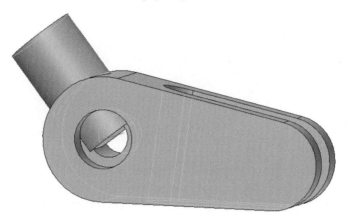

In AutoCAD, the solution to shortening the peg would be to try to position a cylinder, and then subtract it. In Inventor, the solution is to rewrite history: in the Model browser, drag **Hole1** to below **Extrusion3** (peg). Notice that the hole now cuts through the peg, making the irritating part go away.

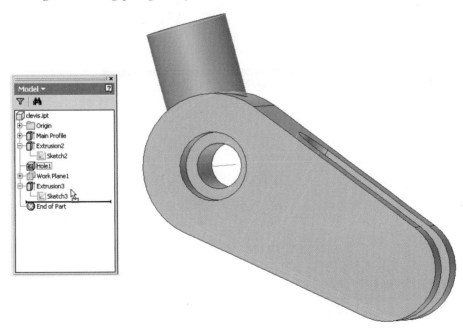

You can drag things around the Model browser, but to a limited extent. If the work axis were drawn relative to the hole, you would not have been able to fix the peg problem by simply dragging it down the history tree (Model browser). In this case, the axis (and hence the work plane, and hence the peg) cannot exist until the hole exists. Descendents cannot exit before ancestors.

Stage 6: Peg End

The end of the peg has flats. These are created by sketching a pair of lines, and then cut-extruding them along the peg by 38mm.

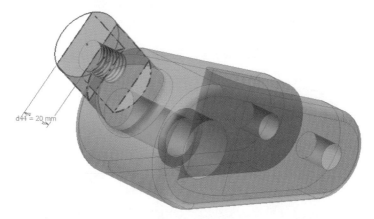

1. Use the **Look At** command to bring the model around to see the end of the peg.

2. Click **Sketch**, and then choose the round end of the peg as the sketch surface. Notice how the circle of the peg is automatically projected onto the sketch plane.

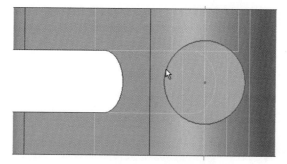

3. Draw a pair of horizontal lines with the **Line** command.

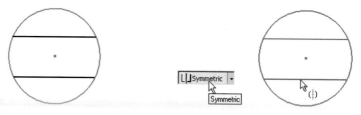

Left: Drawing two parallel lines.
Right: Assigning Symmetric constraints to both lines.

4. Use the **Symmetric** constraint to center them. As the technical editor notes, this is not Inventor; this is geometry. If the two lines are symmetrical about the center of the circle, then they are forced to be the same length. (Alternatively, if the are constrained to be equal length, then they will also be symmetrical.)

5. Use **General Dimension** to specify the distance between the lines, **20**mm.

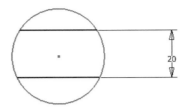

6. The sketch is complete; exit sketch mode by clicking **Return**.

   Rotate the view in anticipation of cutting the peg.

7. Use the **Extrude** command to cut away the sides of the peg, as follows:

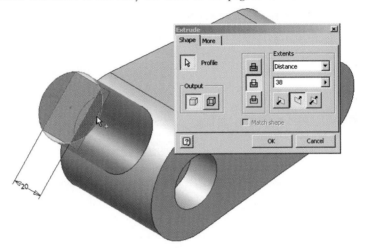

- **Profile** = select the two crescent shapes, shown in blue and red, above. As you move the cursor over a profile candidate, it turns red; when you click to select it, it turns blue.

- **Operation** = Cut

- **Distance** = 38mm

- **Direction** = down

Click **OK** to effect the cuts.

8. Save your work.

> **TIP** *You can preview effects of each Boolean operation by clicking the **Join**, **Cut**, and **Intersect** buttons in the Extrude dialog box.*

Stage 6b: Peg Hole

The peg has a threaded hole located in the center of the flats. To locate the hole accurately, use construction lines.

1. Click **Sketch**, and then pick the face of one of the flats.

2. To draw a construction line, choose **Construction** from the Standard toolbar, and then choose **Line** from the 2D Sketch panel. In Inventor, Construction acts as a modifier for the next drawing command, turning the object into a construction object — which is shown dashed.

3. Draw the construction line from midpoint to midpoint. Notice that Inventor automatically provides midpoint object snap; there is no need to turn it on or off, as in AutoCAD.

Left: *Drawing the construction line.*
Right: *Center point added.*

To locate the center for the Hole command, you'll also need to draw a point centered on the line. For this, use the **Point, Center Point** command.

4. Now you can use the midpoint of the construction line to locate the threaded hole. Click **Return** to return to part mode, and then click **Hole** to specify the hole's parameters:

   • **Placement** = From Sketch

   • **Centers** = point located at the center of the line

- **Termination** = All Through
- **Hole Type** = Tapped
- **Thread Type** = ANSI Metric M Profile
- **Size** = 20

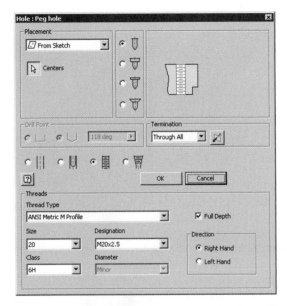

Click **OK** to see the result. Note that the thread is simulated visually.

Stage 6c: Small Hole

While you're at it, add the remaining hole.

1. The small hole at the end of the clevis is **15**mm in diameter. Use Concentric placement.

   You can use the same Hole dialog box to place all holes at the same time, even when they have different parameters.

2. Save your work.

> **TIP** *To edit the hole (or any other feature), select it from the Model browser. Right-click, and then choose* **Edit Feature***. The appropriate editing function or dialog box appears. In Inventor, the same dialog boxes used to create features are also used to edit them.*

Stage 7: Fillets

The clevis has fillets along three edges. They have a radius of 5mm.

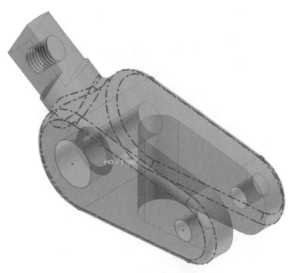

1. From the Part Features panel, choose **Fillets**. Notice the Fillets dialog box.

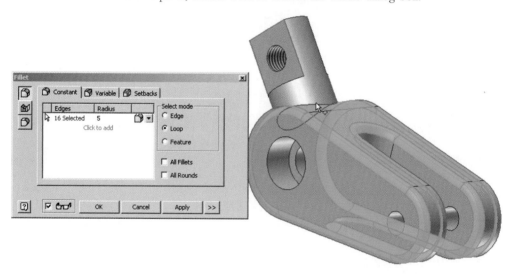

2. Make the following changes to the dialog box's settings:

   • **Radius** = 5mm

   • **Select Mode** = Loop

3. Choose the three loops indicated in the figure above. Notice that Inventor previews them, so that you can tell whether they are correct.

   (The **Click to Add** item allows you to add additional, different fillet radii; these can be applied to individual loops. You control all fillets from a single editing command.)

4. Click **OK**, and Inventor applies the fillets. Save your work.

Stage 8: Material

The clevis is made of brass. Inventor can assign physical properties to parts. These properties allow Inventor to analyze parts and assemblies and check they that will operate as expected — before being built. The last stage in this tutorial is to assign the material, as follows:

1. From the **File** menu, choose **Properties**.

2. In the Properties dialog box, click the **Physical** tab.

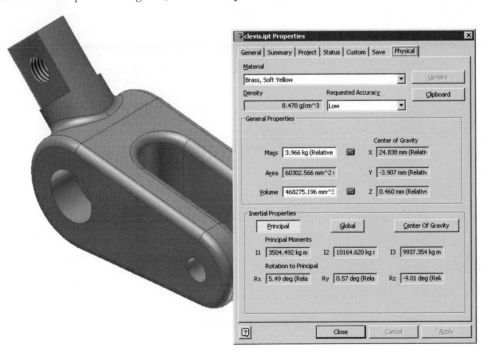

3. From the **Material** droplist, choose "Brass." (You can define custom materials in the Format | Style and Standard Editor dialog box.)

4. Click **Apply**.

5. Click **Close** to exit the dialog box. Notice that the part changes from generic gray to shiny brass.

(The **As Material** droplist on the Standard toolbar only applies colors and textures to parts, in the manner that AutoCAD does. You have to use the Properties dialog box to assign real-world materials. The droplist could be used to simulate gold-plated steel, for example. The technical editor's favorite material is Green Meteorite, which was named "kryptonite" in earlier releases.)

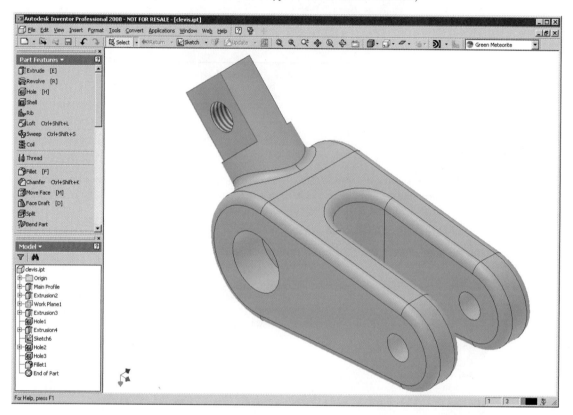

Summary

This chapter showed you how to create a part in Inventor, the building block from which complex assemblies — entire machines — are created. As you saw, it all begins with a simple 2D sketch.

In the next chapter, you go through a similar process, except that you begin with a drawing brought over from AutoCAD.

Chapter 5

AutoCAD to Inventor Tutorial

In the last tutorial, you created a 3D model from scratch in Inventor. In this tutorial, you instead begin with a 2D AutoCAD drawing, paste it into an Inventor drawing, and then convert it to a 3D part. After the correct conditions are set up, you'll find that this process proceeds quickly.

The stages in this tutorial are as follows:

1. In AutoCAD's model tab, copy appropriate portions of a 2D drawing to the Clipboard.
2. In Inventor, paste the drawing into the 2D Sketch environment.
3. Specify the base view.
4. Create the projected views.
5. Extrude the three views to create the 3D part.

> **TIP** *This tutorial requires that you download the "2D to 3D Tool for Inventor" from the Autodesk Labs Web site, as described in the boxed text on the next page.*

IN THIS CHAPTER

- Converting AutoCAD drawings to Inventor through the Clipboard.
- Turning 2D sketches into 3D parts semiautomatically.
- Using the join, cut, and intersect modes of the Extrude command.

2D-TO-3D TOOL SUMMARY

Installing 2D To 3D Tool

For this tutorial, the 2D To 3D Tool must be download from the Autodesk Labs Web site, and then installed on your computer. Close Inventor, and then follow these steps:

1. With your Web browser, go to <u>labs.autodesk.com</u>, and then look for the **2D to 3D Tool for Inventor** link.

2. Click the **Download** button, and then agree to the Terms of Use by scrolling through all the text.

3. Click **Download** again. The download is a Zip file. (If the download fails, you may have to use Firefox or Internet Explorer.)

4. When the Web browser asks, choose **Open**, and then run *ai2dto3d_setup.exe*.

5. Follow the instructions of Setup, allowing it to install the *.dll* file where it wants.

The next time you start Inventor, the 2D To 3D panel appears in the 2D Sketch panel's droplist.

2D To 3D Tool Commands

The tool set is brief, consisting of these commands:

Base View — starts the conversion process by displaying a 3D cube, and prompting you to choose a face on the cube, followed by choosing the 2D elements for the base view.

Projected View — prompts you to choose the elements for the side and top views; places them there automatically.

Sketch Alignment — aligns sketches as front, top, and side views, if necessary.

Extrude — converts the 2D sketches into 3D features.

2D To 3D Tool Tips

-   The 2D sketch should contain only those objects needed by the extrusion process.

-   The resulting 3D model is only as accurate as the 2D sketch; the technical editor warns you to check for cheated dimensions in the source drawings.

-   Ensure critical details are the correct size and in the correct locations.

-   Discard or ignore unnecessary objects, such as title blocks, construction lines, hidden lines, and text.

-   The three views (front, top, and side) of the sketch should line up; if they do not, use the Sketch Alignment command.

-   Only the first extrusion is join (union); the others usually cut (subtract) or intersect the first one.

In this tutorial, you convert a 2D AutoCAD drawing into a 3D Inventor part. Both are illustrated below.

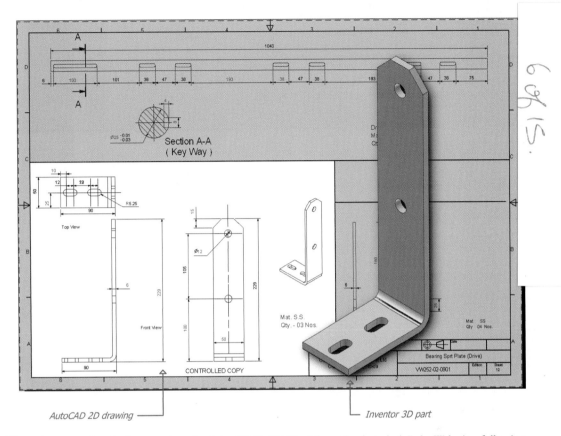

AutoCAD 2D drawing ⎯⎯⎯⎯⎯⎯⎯   ⎯ Inventor 3D part

If you have not already done so, download the 2D To 3D Tool from the Autodesk Labs Web site, following the instructions on the facing page.

Stage 1: Copy Drawing from AutoCAD

In the first stage of the tutorial, you copy the 2D drawing in AutoCAD to the Clipboard.

1. Start AutoCAD, and then open the *VW252-02-0901.dwg* sample drawing. You can find it in AutoCAD 2008's *\Sample\Sheet Sets\Manufacturing* folder.

2. If necessary, click the **Model** tab to switch to model space, because the objects you need are in model space.

3. Using the cursor, choose the objects shown in red. See figure below. The only objects you need are those making up the outline of the three shapes and the holes. If you choose unneeded objects, hold down the **Shift** key, and then select them to remove them from the selection set.

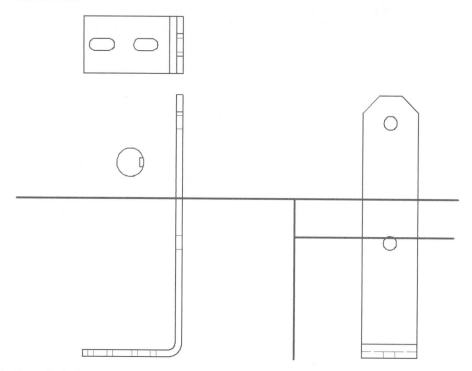

4. Press **Ctrl+C**, and then wait a moment as AutoCAD copies the objects to the Windows Clipboard.

   When the 'Command:' prompt returns, you know that the objects are now stored in the Clipboard.

   > **TIP** *Be careful that you do not copy anything else to the Clipboard between now and step 3 below; otherwise the process fails. If it does fail, return to AutoCAD and recopy the objects to the Clipboard with* ***Ctrl+C***.

Stage 2: Paste Drawing into Inventor

In this stage of the tutorial, you import the AutoCAD drawing into Inventor by pasting it from the Clipboard.

1. Start Inventor by double-clicking the icon on your computer's desktop.

Autodesk Inventor
Professional 2008

2. (If necessary, click the **New File** button in the Quick Launch area.) In the New File dialog box, click the **Metric** tab.

Choose "Standard(mm).ipt," and then click **OK**. Notice that Inventor opens a new part file (*Part1.ipt*), and then starts in 2D sketch mode.

3. Press **Ctrl+V** to paste the AutoCAD objects into Inventor.

4. After several moments, a blue dashed rectangle appears. Its purpose is to help you position the pasted data; in this case, however, it matters not where you put the imported objects, so just click anywhere in the drawing area.

   (It *may* matter, rejoins the technical editor, if you plan to extract properties like center of gravity. In this case, import the model, and then project the center point; move all objects from a logical base point to the center point.)

5. If necessary, press **Home** to see the entire imported drawing.

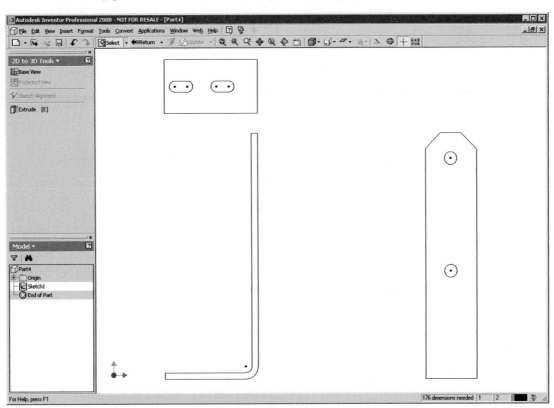

Stage 3: Specify the Base View

In this stage, you start the conversion process by identifying the base view.

1. In the **2D Sketch** panel droplist, choose "2D to 3D Tools." Notice the four commands, which are listed in operational order.

2. Click **Base View**. Notice that Inventor changes the viewpoint to isometric, and then displays a semitransparent cube. The 2D objects will be attached to the cube, and then extruded.

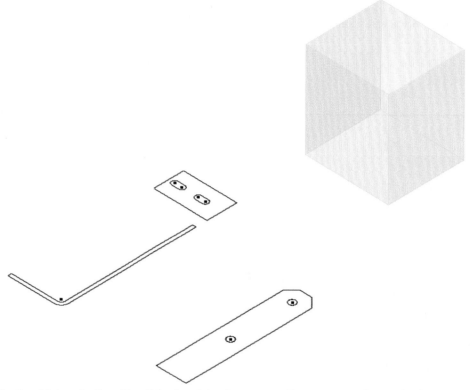

3. In addition, the Base View Selection dialog box appears. It prompts you for two items:

   Select the Face For the Base View — choose one face of the cube. The one you choose becomes the side for the objects selected by the second prompt. For this tutorial, choose the face highlighted in red below.

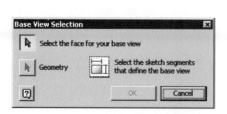

Geometry — choose the geometry to be attached to the selected face. In this case, the ones selected in red.

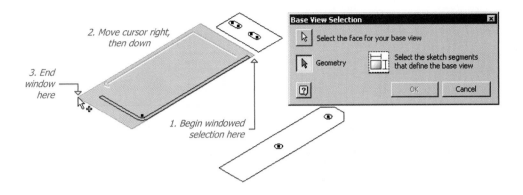

2. Move cursor right, then down

3. End window here

1. Begin windowed selection here

TIP *It can be tricky doing windowed selections in isometric view. I suggest starting at the right corner, and then moving left, as shown above.*

4. Click **OK**.

Notice that Inventor copies the selected geometry onto the chosen face, and then the cube disappears.

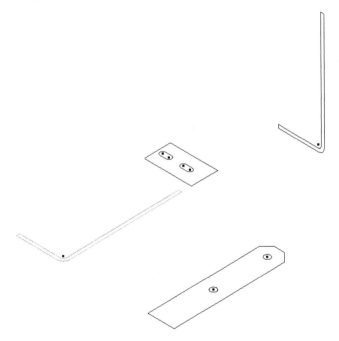

Stage 4: Specify the Projected Views

In this stage, Inventor adds the remaining objects to the remaining faces. Your only task is to choose the necessary objects.

1. In the panel, click **Projected Views**. Notice that the cube reappears, this time elongated to fit the base view objects.

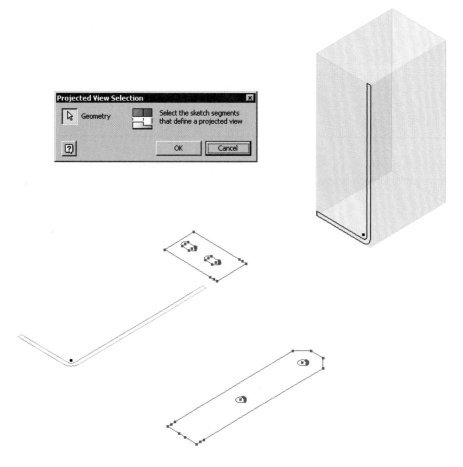

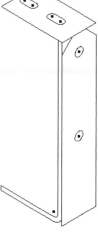

2. Choose the remaining objects — all of them. Inventor automatically figures out where they belong by their relationship to the base view.

   Be very careful to not select any extra objects or previously selected objects, warns the technical editor. This can confuse the alignment process.

3. Click **OK**. Notice that copies appear where the rectangular box once stood.

4. If necessary, use the **Z** (Zoom Window) shortcut to make the 3D entity larger.

With the objects positioned correctly in 3D space, the final stage is to extrude each sketch to create the 3D part.

Stage 5: Extrude the 2D Sketches

In the final stage, you extrude each 2D sketch to create the 3D part.

1. In the panel, choose **Extrude**. Notice the Extrude dialog box.

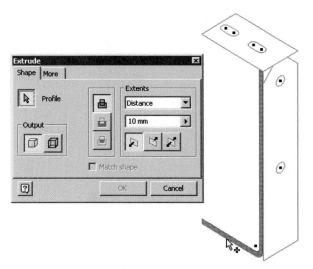

2. Pick a profile.

   You can begin with any of the three profiles (2D sketches). Sometimes it makes more sense to start with a specific profile, but in this case any of these three works just fine.

3. Specify the distance, as follows:

   a. In the distance droplist, click the ▶ arrow button. Notice the flyout.

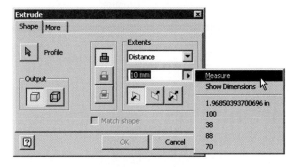

   b. Choose **Measure**. Notice the cursor now has a ruler icon.

   c. Pick an object that represents the depth, such as the line shown in red, above. Notice that the extrusion jumps out by that distance.

4. Reverse the direction of the extrusion by clicking the  button.

Click **OK**.

5. The first extrusion is in place. The remaining two extrusions shape the first one, like this:

   a. Click the **Extrude** button.

   b. Choose the vertical profile. Notice that it turns red.

   c. To specify the thickness of the extrusion, click the **Distance** droplist, and then choose "To." This option measures the thicknesses *to* another face.

d. Move the cursor over the face shown outlined in red below, and then click.

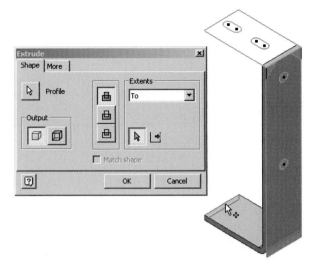

Notice that the extrusion is created, but it's not quite right.

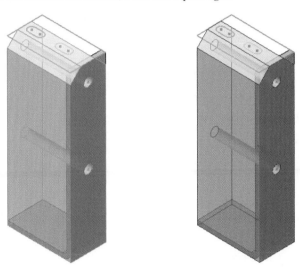

Left: *Join (union) extrusion.*
Right: *Intersect extrusion.*

e. This extrusion must intersect with the first one, to cut away unneeded portions, specifically the two holes and chamfers. Click the   **Intersect** button.

f. Click **OK**.

6. Apply the **Extrude** command to the remaining profile, as follows:

   a. With this profile, you need only punch the two holes through the base; select just the holes, not the entire profile, as illustrated below.

   b. For the distance (thickness), choose the **All** option. Inventor automatically cuts the slots all the way through.

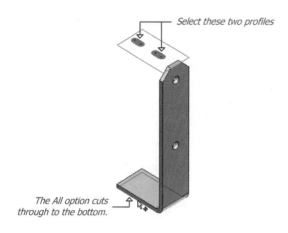

Select these two profiles

The All option cuts through to the bottom.

Left: *Selecting profiles .*
Right: *Subtracting the holes from the part.*

   c. Use **Cut** (subtract) mode to remove the holes from the part.

7. Click **OK**. The holes appear, and the part is completed.

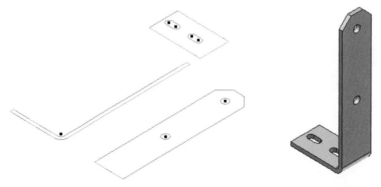

8. Clean up the drawing by erasing the 2D sketches, which are no longer needed. The sketches are shown in gray at left, above.

   Remember to save your work.

Summary

This tutorial shows how quickly 2D sketches from AutoCAD and Inventor can be converted into 3D parts.

In the next chapter, the final tutorial of this series shows you how to combine parts into assemblies.

Chapter 6

Parts to Assemblies Tutorial

The primary purpose of Inventor is to create assemblies, large and small — bottling plants, can openers, and earth moving machinery. (For ideas on how Inventor is used, go to www.autodesk.com/inventor, and then click on Customer Stories.) All assemblies are made of parts, from as few as two to several thousand.

The previous tutorials showed you how to create parts. In this tutorial, you combine parts to create the assembly illustrated on the next page. (Because it can only design parts, Inventor LT is unable to make the assembly in this tutorial.)

To create assemblies, you just repeat the same two steps to add each part:

1. Use the **Place Component** command to insert a part into the assembly 3D model.

2. Use the **Constraint** command to apply constraints that fix the part in place.

That's all! The tricky part is knowing which constraint to use, and how. Thus this tutorial employs constraints in several different ways. These are the constraints available in Inventor, listed in order of most common use:

- **Mate** connects the features of two parts.

- **Insert** connects two features along their axes and at a plane; useful for putting bolts in holes.

- **Angle** fixes two features at an angle.

- **Tangent** connects two features at their points of tangency.

IN THIS CHAPTER

- Starting new assemblies.
- Creating project paths.
- Adding parts to assemblies.
- Constraining parts in assemblies with mate, angle, and insert constraints.

ASSEMBLY SUMMARY

The exploded diagram illustrates the parts needed for this chapter's tutorial. The numbers refer to the tutorial's order of assembly. To access these part files, go to www.upfrontezine.com/vise. Download the vise.zip file, and then unzip the contents of the file to a folder on your computer's hard drive.

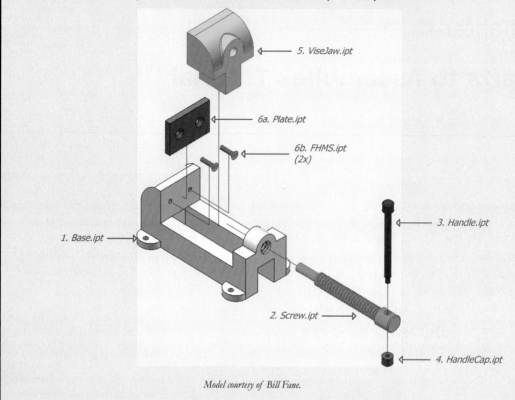

Model courtesy of Bill Fane.

An unconstrained part in space has six degrees of freedom: it can move in the x, y, or z direction and rotate about each of the three axes. Constraints reduce degrees of freedom. For instance, a mate point-to-plane constraint removes one degree of freedom, while an insert constraining removes five.

By reducing degrees of freedom, constraints keep parts together, like the glue used for assembling plastic model kits. Most times, you'll find that parts need three constraints, because that's how many are needed to fix a part in 3D space. Sometimes, you can employ fewer than three by sharing constraints between parts.

Pre-Tutorial Preparation

This tutorial uses part files that you need to download from www.upfrontezine.com/vise. The seven part files are in a Zip file named vise.zip.

1. Use your Web browser to go to www.upfrontezine.com/vise.

2. Click on the download link, and then save the vise.zip file to your computer's hard drive.

3. Unzip the file to a folder on your computer's hard drive.

With the parts downloaded to your computer, you are ready to proceed with this tutorial.

Starting New Assemblies

To create a new assembly, begin by opening a new *.iam* assembly file, like this:

1. Start Inventor.

2. In the New File dialog box, choose an *.iam* template file. For this tutorial, click the **English** tab, and then select "standard(in).iam."

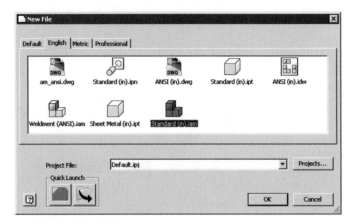

3. But don't press OK just yet. It helps to set a path to this project.

Project Paths

The Project File droplist tells Inventor where to find files. Generally, all files related to a single project are placed in the same folder and subfolders, either on your computer or on one that's networked. *Project* is the name for all files that relate to an assembly: assembly, parts, drawings, presentations, and other files.

Setting up a project file for this tutorial will make it easier for you to access the *.ipt* part files needed by the assembly. To create the project file, follow these steps:

1. In the New File dialog box, click **Projects**. Notice the Projects dialog box.

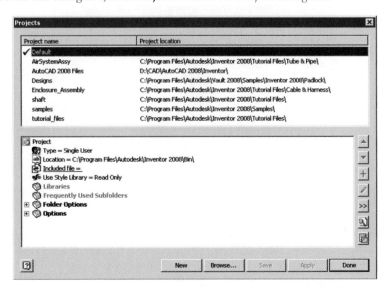

2. To create a new project for this tutorial, click **New**.

Notice the Inventor Project Wizard dialog box.

3. Choose **New Single User Project**, and then click **Next**. (You would choose New Vault Project when two or more work on the same assembly.)

4. Enter a name for the project:

    Name = Vise

    Click the **...** browse button, and then choose the folder that holds the vise part files that you previously downloaded.

    Click **Next**.

5. You won't be using library files in this tutorial, so click **Finish**.

6. Notice that "Vise" is added to the list of project names in the Projects dialog box.

    To change the Project to Vise, click **Apply**, and then click **Done**.

7. Back in the New File dialog box, notice that *Vise.ipj* is listed in the Project File. Each time in this tutorial you need to add another part to the assembly, Inventor will automatically open the Place Component dialog box to the folder defined by this project.

    Click **OK** to open the new assembly drawing.

Notice the blank assembly file, with the Assembly panel of commands.

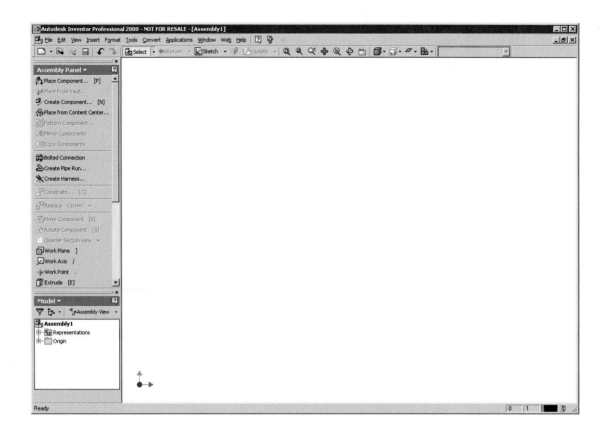

Assembling the Parts

With the project path set and a new assembly drawing opened, it's time to start adding parts.

Stage 1: Base

The first part you add to the assembly is called the "base component." It is the most important one, the foundation upon which all other parts are added. Good practice is to start with the biggest, heaviest lump, and then to add other components in a logical assembly sequence, advises the technical editor.

> **TIP** *Although components can be added in any order (and then reordered by drag-n-drop in the browser), sequence is important for bills of material and part lists. (Their sequences numbers are nominally based on the browser sequence.)*

For this tutorial, the base component is obviously the base of the vise. To place the base, follow these steps:

1. From the Assembly panel, click **Place Component**. (It is called "component" because you can place parts and assemblies.)

    Notice that the Place Component dialog box opens to the folder specified by the *Vise.ipj* project file. See figure on the next page.

    But also notice that the Project File droplist is grayed out, because Inventor assumes that all work you do in a single session relates to a single project. You can open files from other projects, but then you have to navigate your computer's folder structure to get to them.

Castle Learning Resource Centre

Finally, notice that Inventor lists both *.ipt* part and *.iam* assembly files as components in the Files of Type droplist. This indicates that you can use other assemblies in this one — subassemblies, as it were.

2. Select "Base.ipt," and then click **Open**.

3. *Don't click!* Notice that Inventor inserts one copy of the base, and ghosts a second one. If you were to click in the window now, Inventor would add another copy of the base. This is useful when you need to insert rapidly more than one bolt or other repeated parts, but not this time.

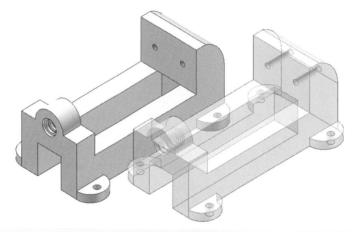

Since you need only one base, press **Esc**. (If you by accident clicked, then remove the second part with the Undo command.)

4. Try to drag the base around the screen. It does not move, because it is *grounded* to the file coordinate system. (Grounding can be removed from, and added to, any component at any time through the right-click menu's **Grounded** toggle.) Usual practice is to ground only the base component — the foundation for all remaining parts.

5. If necessary, use the **Rotate** command to rotate the view, then right-click and use the **Common View** command to set a more useful isometric view.

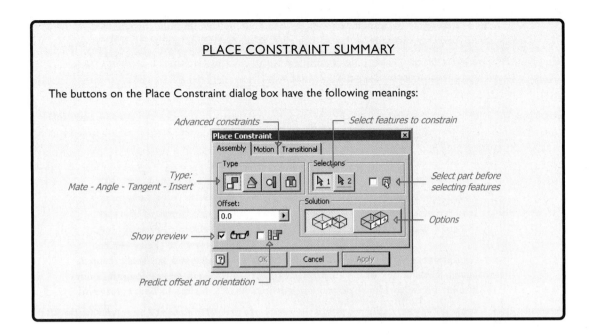

PLACE CONSTRAINT SUMMARY

The buttons on the Place Constraint dialog box have the following meanings:

Advanced constraints

Select features to constrain

Type:
Mate - Angle - Tangent - Insert

Select part before
selecting features

Options

Show preview

Predict offset and orientation

Stage 2: Screw

With the base component in place, it's time to add the next part, the vise's screw drive.

1. Press the spacebar to repeat the Place Component command.

   In the Place Component dialog box, select "screw.ipt," and then click **OK**.

2. Click to place one copy of the screw, and then press **Esc** to stop further placement.

3. Drag the screw around the window and notice that it moves. This shows that you need to apply constraints to keep it fixed in the base component.

4. In the Assembly panel, click **Constraint**. Notice the Place Constraint dialog box. You'll find that these constraints act somewhat differently from the ones you used for 2D sketching in earlier tutorials.

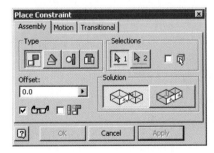

5. Click the **Mate** button. This constraint forces the face of one part to align with the face of another, or one axis to align with another. (It also mates points and combinations, such as a point to a line, a line to a face, and so on.)

6. Notice that the **1** button is depressed. This means that Inventor is waiting for you to specify the first mate — the axis of base's large hole.

This can be a bit tricky. Basically, move the cursor around the base until the long, red centerline appears (as shown below), and then click. If necessary, zoom in for a closer look.

7. After you select the first feature, the depressed **2** button indicates Inventor is ready for you to specify the second mate. Choose axis of screw.

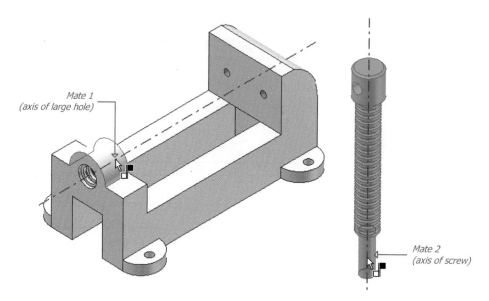

Mate 1
(axis of large hole)

Mate 2
(axis of screw)

8. Click **Apply**. Notice that the screw jumps in line with the hole. That's because the base is fixed; the screw has to adapt itself to the base component.

9. Now drag the screw. Notice that you can move it back and forth along the axis, and rotate it about the axis. But you cannot move it away from the base. These motions indicate that the screw needs two more constraints to keep it in place: one to prevent it from moving back and forth, the other to keep it from rotating.

10. You can reuse the Place Constraint dialog box for the next constraint. Enter these options:

   a. Select the flat face on the end of the base.

   b. Use the **Rotate** command to see the flat annular (ring-shaped) face on the underside of the head of the screw. Press **Esc** to exit view-rotation mode.

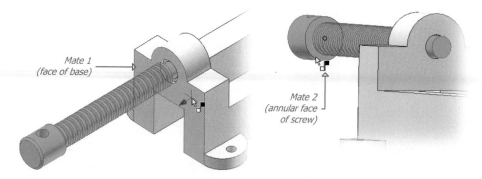

Mate 1
(face of base)

Mate 2
(annular face
of screw)

   c. Select the annular face of the screw, as illustrated above. (The green dot indicates the centre of the circular edge.)

(Alternatively, hover the cursor over the location, and then use the 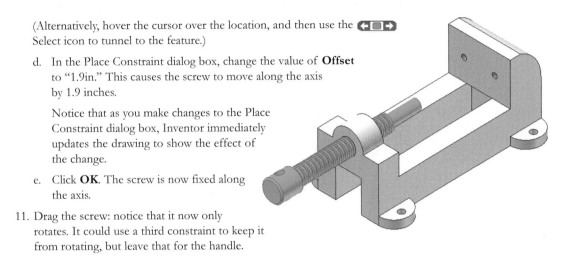 Select icon to tunnel to the feature.)

d. In the Place Constraint dialog box, change the value of **Offset** to "1.9in." This causes the screw to move along the axis by 1.9 inches.

Notice that as you make changes to the Place Constraint dialog box, Inventor immediately updates the drawing to show the effect of the change.

e. Click **OK**. The screw is now fixed along the axis.

11. Drag the screw: notice that it now only rotates. It could use a third constraint to keep it from rotating, but leave that for the handle.

The technical editor comments: this stage of the tutorial used two mate constraints — one line-to-line, and one face-to-face — to remove five degrees of freedom. (One rotation degree remains.) Optionally, you could use an Insert constraint, because it removes five degrees of freedom. If you did, you would select two circular edges, such as the edge of a hole and the edge of the head. This single constraint replaces two Mate constraints. (Some competitors to Inventor lack the Insert constraint, and require the use of the two Mate constraints.)

Stage 3: Handle

Attach the handle to the screw, like this:

1. Choose the **Place Component**, select "handle.ipt," and then click **Open**.

2. Click once to place one copy of the handle, and then press **Esc**.

3. Choose **Constraint**, and then use the **Mate** constraint to link the axis of handle with the axis of the hole at the end of the screw, as shown by the figure.

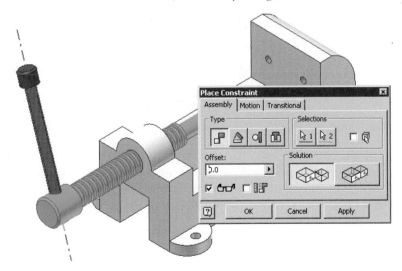

4. Click **OK**.

5. Drag handle; notice that the action rotates the screw. This shows that applying a constraint to the handle also applies to the screw.

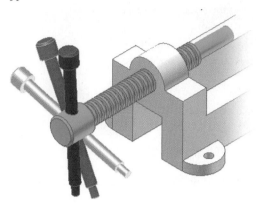

6. Press spacebar to reopen the Place Constraint dialog. Click **Angle** to place an angle constraint between the centre line of the handle and any edge of the base. This removes the rotation freedom of the screw by constraining the handle.

7. Click **OK**.

Stage 4: HandleCap

The next step is to add the end cap to the handle.

1. Use the **Place Component** command to add one copy of the "handelcap.ipt" part file to the assembly drawing. (You may want to use the mouse wheel to zoom into the cap/handle area.)

2. Start the **Constraint** command, and then follow these steps:

   - Click the **Insert** button. Notice that parts of the dialog box change to accommodate the Insert Constraint's options.

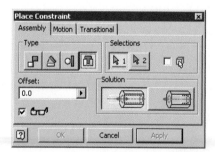

Insert constraints always require circular or arcuate (arc shaped) edges, but never centre lines. (It may seem to involve center lines, because both the circular edge and the centre line are highlighted when the circular edge is selected.)

- For **1**, choose the circular face at the bottom of the handle.
- For **2**, choose the inside circular face inside the cap.

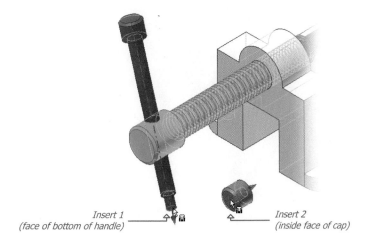

Insert 1
(face of bottom of handle) ——— *Insert 2*
(inside face of cap)

3. Click **OK**, and then press **Home** to zoom out, if necessary.

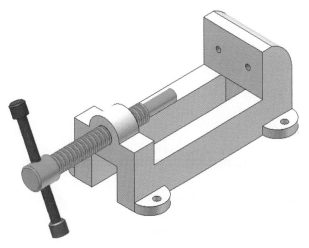

4. Drag the cap to check whether it moves with the handle and does not detach: the cap is fully constrained, even if the handle is not.

5. You constrain the handle indirectly, by specifying the distance from the screw to the cap, like this:

    a. Click **Constraint**.

b. For feature **1**, choose the end of the cap, as illustrated below.

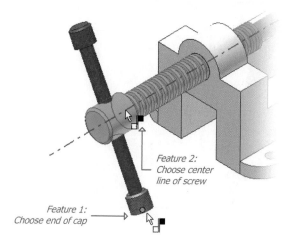

Feature 2:
Choose center
line of screw

Feature 1:
Choose end of cap

c. For feature **2**, choose the centerline of the screw.

d. Specify an **Offset** distance of about **2** inches. As you enter the value, notice that Inventor immediately updates the assembly.

e. Click **OK**. The handle no longer slides back and forth.

TIPS Constraint offsets (distances) need not be fixed at 0 or 1.9 or 2 inches. To change the value, follow these steps:

1. In the Model browser, open a part, such as **HandleCap:1**.
2. Click a constraint name, such as **Mate(2.000in)**. Notice the edit box that opens at the bottom of the Model browser:
3. Change the value, and then press **Enter**. Notice that the handle changes its position.

To display the degrees of freedom (DOF) on parts, chose **Degrees of Freedom** from the **View** menu, or press **Ctrl+Shift+E**. As you apply constraints, portions of the DOF symbols gradually disappear, showing the remaining DOFs.

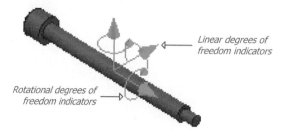

Linear degrees of
freedom indicators

Rotational degrees of
freedom indicators

Stage 5: ViseJaw

The vise's jaw gets placed next.

1. Use the **Place Component** command to insert one copy of *vicejaw.ipt*.

2. Use the **Constraint** command's **Mate** option to position the jaw on the rails of the base, as illustrated below.

3. When you've selected the faces, click **Apply**.

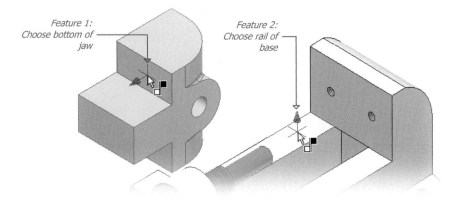

Feature 1:
Choose bottom of jaw

Feature 2:
Choose rail of base

4. With the Place Constraint dialog box still open, click the **Insert** option to align the jaw with the screw, as illustrated below. The jaw jumps nicely into place. Click **OK** when done.

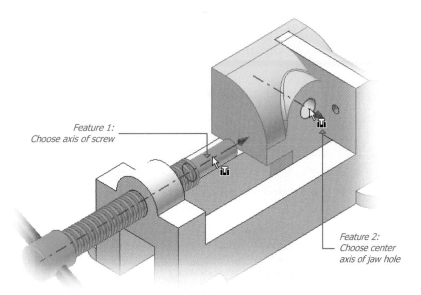

Feature 1:
Choose axis of screw

Feature 2:
Choose center axis of jaw hole

5. Attempt to drag the jaw. Notice that it does not move; it is fully constrained.

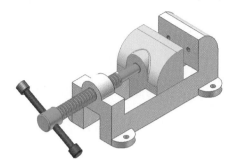

Stage 6: Plate & FHMS

The plate is a two-fer. You need to attach the plate to the base with two screws. The approach is to add the two machine screws to the plate, creating a form of "subassembly." (A true subassembly is an assembly that's added to another one.) Then you attach the plate to the base. Here's how:

1. Use the **Place Component** command to insert one copy of *plate.ipt* and two of *fhms.ipt* in the 3D model.

2. To see both sides of the parts more easily, select **Hidden Edge Display** from the Standard toolbar.

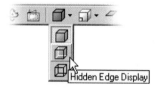

3. Use the **Constraint** command's **Mate** option to attach the machines screws snugly to the plate. The trick is to select the countersink faces on each part. And the trick to accomplishing this is as follows:

    a. First, turn on the ![icon] **Pick Part First** option in the Place Constraint dialog box.

    b. Pick the plate.

    c. Move the cursor over the plate until Inventor finds the countersink feature. (Hover the cursor over the countersink feature until the ![icon] Select icon appears. Click its arrows until Inventor highlights the countersink, and then click the green square.)

    Make sure the countersink is selected; it should be the only feature displayed in red.

d. For **2** choose the countersink feature on one of the machine screws. Click **Apply**.

e. Repeat to place the second machine screw. Notice that it also jumps into position.

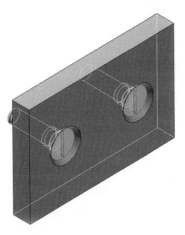

4. Drag the machines screws. Notice that you can rotate them. To fix them in place, use the **Constraint** command's **Angle** option.

a. For feature **1**, select and edge of one machine screw's slot, as illustrated below. Notice the red direction vector.

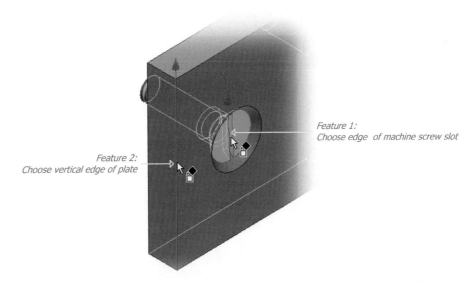

Feature 1:
Choose edge of machine screw slot

Feature 2:
Choose vertical edge of plate

b. Click **2**, and then choose one of the plate's vertical edges. Click **Apply**.

c. Repeat for the second screw, and then click OK. When you attempt to drag the machine screws, they no longer move.

5. Now attach the plate to the base, like this:

   a. Press the spacebar, and then click the **Insert** button. (Insert constraints can also constrain holes to holes.)

   b. Select the circular edge of one machine screw hole in the base, and of one plate hole. You may find it easier to use the 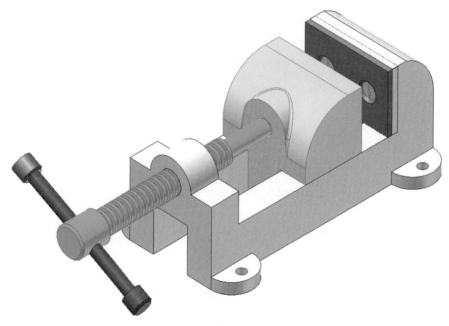 **Pick Part First** option.

   c. When the plate attaches itself to the base, the machine screws might be facing the wrong direction. If so, click the **Aligned** button to reverse the direction.

   c. Click **OK**, and then drag the plate. It rotates, meaning that you still need to constrain it further.

   d. Repeat the above steps for the other screw to fix everything in place. Alternatively, apply an Angle constraint between a base edge and a plate edge.

This concludes the tutorial.

Summary

Congratulations! The assembly is complete. Save your work, and print off a copy to show proudly to family and friends. As a former AutoCAD user, you've got a lot of this Inventor program figured out.

The next section of this book delves into the details of importing AutoCAD drawings into Inventor – and exporting them back again.

Part III

Viewing, Importing, and Exporting DWG Files

Chapter 7

Working with AutoCAD Drawings in Inventor

As an AutoCAD user, you'll want to reuse your collection of drawings (*.dwg* files) in Inventor. The good news is that Inventor reads AutoCAD's native file format, DWG.

In the technical editor's experience with AutoCAD and Inventor, he suggests that there are four cases that cover the bulk of needs for reusing AutoCAD drawings in Inventor:

- Reusing AutoCAD 2D drawings as sketch geometry in Inventor parts. They are already drawn in AutoCAD, and so there is no need to draw them again in Inventor.

- Reusing AutoCAD 2D drawings of title blocks, borders, annotations with Inventor drawings. No need to draw them again, especially when a consistent appearance is needed. For instance, as soon as contractors open AutoCAD drawings in Inventor, they have drawings that match the customers' title blocks.

- Reusing AutoCAD 3D drawings as Inventor parts. Many users dabble with AutoCAD 3D before switching to Inventor and don't want to throw away the parts (3D models) they've already built.

- Reusing Inventor 2D drawings (either type) in AutoCAD for interoperability with other AutoCAD users.

The other possibilities — such as redefining AutoCAD blocks, round-tripping, and so on — are not as common, but can be done with Inventor.

IN THIS CHAPTER

- Understanding how Inventor's DWG format differs.
- Opening AutoCAD drawings for viewing
- Viewing, measuring, printing, and adding to AutoCAD drawings.
- Changing layer properties.
- Converting (importing) AutoCAD drawings.

Primary Support for 2D

Inventor's support for AutoCAD allows you to reuse the content you've already created in AutoCAD, whether drawings or blocks or title blocks. You can use Inventor to reconstruct AutoCAD's 2D drawings as 3D parts. Or, you can use Inventor to inspect AutoCAD drawings, such as to retrieve measurements.

Autodesk's philosophy is to initially support 2D AutoCAD drawings. That's why most of the options in Inventor and most of the examples in the remaining chapters of this book are 2D. There is some support for 3D, but it is not as well developed as 2D.

There is a problem, however: AutoCAD and Inventor have two different ways of looking at CAD data, and so have a certain level of incompatibility. Autodesk has worked on solving this problem, and has come up with two solutions: **view** or **import**.

Here's the difference between the two solutions:

View

View AutoCAD drawings in Inventor

Pros:

> Process is transparent; there are no conversion options to work through.
>
> AutoCAD drawings look 100% accurate in Inventor, with a few exceptions.
>
> Annotative objects can be added by Inventor and opened in AutoCAD, but not edited.

Con:

> Objects in AutoCAD drawings cannot be edited by Inventor, with a few exceptions.

Import

Import AutoCAD drawings to Inventor

Pros:

> Imported AutoCAD objects can be fully edited by Inventor.
>
> Imported drawings can be opened back in AutoCAD, but not edited.

Cons:

> A few AutoCAD objects are not imported with 100% accuracy.
>
> Users need to make decisions for optimized conversion processes.

Thus, you have two ways to go:

- **Open** your AutoCAD drawings for viewing (and annotative additions) in Inventor; best for round-tripping drawings between AutoCAD and Inventor.

- **Import** your AutoCAD drawings for full editing in Inventor; best for using AutoCAD drawings as the basis of new Inventor 3D models.

DWG: BEHIND THE SCENES

Behind the scenes, Inventor uses Autodesk's DWG TrueConnect technology to read AutoCAD data directly.

Inventor uses the same DWG format as AutoCAD. The difference is that Inventor writes additional data to custom objects in the *.dwg* file in the same way MDT, AutoCAD Mechanical, and others do. Object Enablers are shipped with AutoCAD 2008 to make the data accessible in AutoCAD, although that accessibility is currently limited.

Autodesk added the "AutoCAD Drawing (*.dwg)" and "Inventor Drawing (*.dwg)" file filters to Inventor as a convenient way to identify those DWG files with and without Inventor data.

The subject is complex enough that I've split it into three chapters: this introductory one, a second one on viewing AutoCAD drawings, and a third on importing them. In summary, the chapters cover these topics:

Chapter 7: Working with AutoCAD Drawings in Inventor

An introduction to opening and converting drawings in Inventor:

> **Open** an AutoCAD drawing in Inventor, and then add to it.

> **Convert** an AutoCAD drawing to Inventor.

Chapter 8: Viewing AutoCAD Drawings with Inventor

In-depth description of viewing drawings in AutoCAD and Inventor:

> **Viewing AutoCAD Drawings in Inventor** — Inventor directly opens *.dwg* files in its Drawing environment with the Open command. AutoCAD data is preserved as read-only AutoCAD data. You can add to the drawing, but you cannot edit any of the objects created by AutoCAD (with a few exceptions).

> **Viewing Inventor Drawings in AutoCAD** — Inventor can save its drawings in DWG format, which can be opened by AutoCAD. Some edits can be made, and then the drawing saved and reopened in Inventor.

Chapter 9: Importing AutoCAD Drawings to Inventor

In-depth coverage of importing drawings between AutoCAD and Inventor:

> **Importing AutoCAD Drawings to Inventor** — Inventor imports *.dwg* files in its Sketch environment with the Insert AutoCAD File command. All of AutoCAD's data are translated into Inventor's format and are fully editable; however, the translation is not perfect.

> **Saving Inventor Drawings to AutoCAD** — Inventor can save its drawings in Inventor DWG format, which can be opened by AutoCAD for viewing and minor editing. (This works only in the Drawing environment, and not from Assembly, Part, Presentation, or any other environment.)

Inventor's DWG File

As an AutoCAD user, you are very familiar with *.dwg* files. The good news is that Inventor reads and writes *.dwg* files; the bad news is that there are now two types of *.dwg* file, which can make things confusing.

In short, the *.dwg* file created by Inventor adds data that cannot be fully edited in AutoCAD. This group of three chapters describes the differences in detail.

This level of *.dwg* compatibility operates best between AutoCAD 2008 and Inventor 2008; a more limited form of interaction is possible between AutoCAD 2007 and Inventor Release 11, and earlier releases. I anticipate that Autodesk will continue to improve the ability of these two programs to exchange data with each other, until it becomes effortless.

(Over the decades, I've collected many 2D and 3D drawing files from most releases of AutoCAD. The oldest 3D solid models in my collection were created by the Advanced Modeling Extension in AutoCAD Release 12. I opened one of them, *steering.dwg*, in Inventor. The drawing opened and displayed correctly in Inventor. By rotating the model, I could tell that it was fully 3D — and not just a 2D representation. This shows that Inventor can read solid model drawings as early as ones created by AutoCAD Release 12.)

Opening AutoCAD Drawings for Viewing

To give you a feel for using AutoCAD drawings in Inventor, this tutorial has you open an AutoCAD 2D sample drawing in Inventor for viewing. You use Inventor to add some annotations, save the drawing, and then open it in AutoCAD. This tutorial is a brief overview; Chapter 8 repeats these tasks in greater detail.

1. Start Inventor.

   From the **File** menu, choose **Open**.

2. In the Open dialog box, change **Files of Type** to "AutoCAD Drawing (*.dwg)."

   In the **Look In** droplist, go to AutoCAD 2008's *help**buildyourworld* folder on whichever drive AutoCAD is installed.

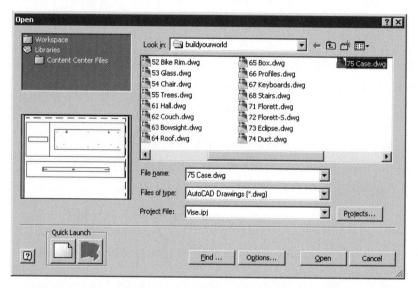

3. Choose "75 Case.dwg" and then click **Open**. (If a dialog box appears and complains about projects, ignore it by clicking **OK**.)

4. Notice that the drawing opens in either in Inventor's Drawing Review or Drawing environment — depending on whether the last AutoCAD view was saved in model or layout tab:

AutoCAD Drawing	Inventor Environment	Functions Available
Saved in Model tab.	Opens in Drawing Review environment.	View, measure, plot.
Saved in Layout tab.	Opens in Drawing environment.	Add viewports, annotate, dimension, change properties.

When an AutoCAD drawing has no viewports in layout tabs, then the Drawing environment will be blank. Should you want to work in Inventor's Drawing environment, you'll need to use AutoCAD to create at least one viewport in a layout tab before opening the drawing in Inventor.

Inventor's Drawing Review environment cannot edit the drawing, neither by changing objects in the AutoCAD drawing nor by adding Inventor objects.

5. If the drawing opens in the Drawing environment, switch to the Drawing Review environment for the purpose of this tutorial. Here's how:

   a. In Model panel, right-click **Model(AutoCAD)**.

b. In the shortcut menu, choose **Activate**.

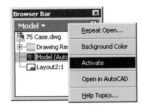

Notice that the environment changes; see figure below.

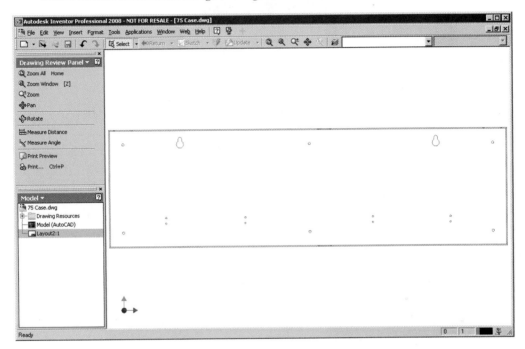

> **TIP** *To change the background color, right-click **Model (AutoCAD)**, and then choose **Background Color** from the shortcut menu. When the Color dialog box appears, choose another color, such as white, and then click **OK**.*

This change affects every AutoCAD drawing opened from now on, until you change the color again.

Tutorial: Viewing AutoCAD Drawings

When AutoCAD drawings are opened for viewing in Inventor's Drawing Review environment, you can measure objects, change the view, and print the drawings. The Drawing Review control panel provides access to the following groups of functions:

Viewing — zoom, pan, and rotate.

Measuring — distances and angles.

Printing — to printers.

1. Try the zoom and pan commands.

   I like pressing the **Home** (zoom all) and **Z** (zoom window) buttons as shortcuts. Alternatively, rotate the mouse scrollwheel to zoom in and out.

   The shortcut for panning is to use the cursor (arrow) keys.

2. Use the **Rotate** command, and notice that it is very similar to AutoCAD's 3dOrbit command. (Curiously, the Rotate command is on the panel, but not in the View menu.)

Tutorial: Measuring AutoCAD Drawings

Follow these steps to measure distances and angles in AutoCAD drawings:

1. In the Drawing Review panel, click **Measure Distance**. Notice that the Measure Distance dialog box appears, and that the cursor has a ruler icon.

2. Pass the cursor over objects in the drawing. As you do, notice that they turn red.

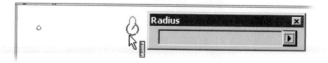

As Inventor recognizes an object, the object turns red, and dialog box's title bar changes:

- **Length** for lines or edges.
- **Diameter** for circles; in addition, a diameter line appears.
- **Radius** for arcs; in addition, a radius line appears, as illustrated by the figure above.

It is possible to measure distances between objects, but sometimes it works for me, and other times it doesn't. Autodesk reports that measuring is meant for 2D only, and so may not work correctly with 3D data.

3. Click the object, and the measurement dialog box reports the length, diameter, or radius.

4. To set units or change the precision, click the 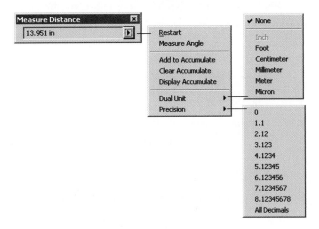 arrow button to see the menus.

The menu provides access to the *accumulate* (addition) mode, as well as a means to switch between measuring angles and distances.

5. To measure angles, choose **Measure Angle**, and then select two objects. Notice that Inventor draws an arc between the two objects, and then reports the angle in the dialog box. Angles can be reported in degrees or radians.

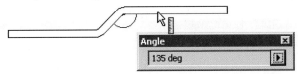

Sophisticated measurements, such as mass and center of gravity, are not available.

Printing AutoCAD Drawings

To print drawings, choose **Print** in the Drawing Review panel. The Print Drawing dialog box provides some options. Note that "Current Sheet" refers to the current Design Review environment, while "All Sheets" includes this environment and the layouts.

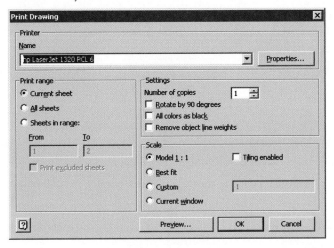

Tutorial: Adding to AutoCAD Drawings in Inventor

You can add Inventor objects to the AutoCAD drawing, but not in the current Drawing Review environment. You need to switch to layout mode, called "sheet" mode or Drawing environment. Here's how:

1. Right-click **Layout:1** (or the name of a layout).

2. From the shortcut menu, choose **Activate**.

Notice that the drawing appears in Inventor's Drawing environment, with its distinctively colored "sheet."

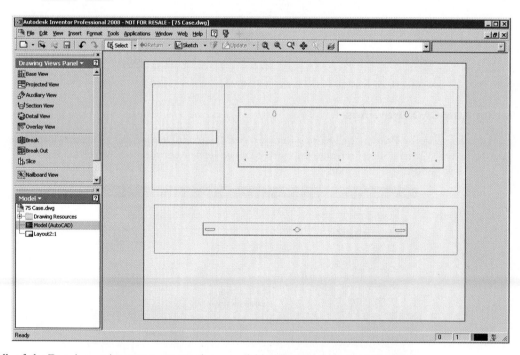

All of the Drawing environment commands are available. These include the following:

- Insert view (viewports) of other Inventor parts, assemblies, and presentations.
- Add text, leaders, and dimensions, as well as tables, balloons, and other annotative symbols.
- Change layer properties.

Tutorial: Adding Views to AutoCAD Drawings

You can add viewports (views) of other Inventor drawings to this AutoCAD drawing, but with a limitation: you cannot place views of this or other AutoCAD drawings. To insert viewports, follow these steps:

1. From the panel, choose **Base View**.

2. In the Drawing View dialog box, click the **Explore Directories** button. (If other drawings are open in Inventor, you'll be able to choose them from the **File** droplist.)

3. From the Open dialog box, choose an Inventor part (**.ipt*), assembly (**.iam*), or presentation (**.ipn*) file, and then click **Open**.

   For this tutorial, choose one of the simple 3D models provided with Inventor, such as *views-6.ipt* found in Inventor 2008's \ *Tutorial Files* folder.

4. Follow the usual steps for placing views in drawings, as described in earlier chapters.

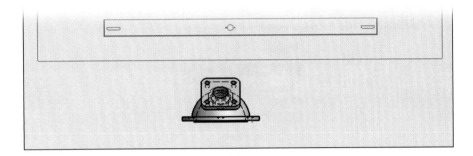

Tutorial: Adding Annotations to AutoCAD Drawings

To add text and other annotation, follow these steps:

1. In the panel, click **Drawing Views Panel**.

2. From the droplist, choose **Drawing Annotation Panel**. Notice that the content of the panel changes to show dimension, text, tables, and other annotative commands.

3. Follow the usual steps for adding dimensions and other annotations to the drawing. When dimensioning, move the cursor over elements, and then wait for the green dot to appear. This means that Inventor has recognized the element.

   (In 3D drawings, Inventor might create incorrect dimensions, because Autodesk intends that only 2D drawings be worked on.)

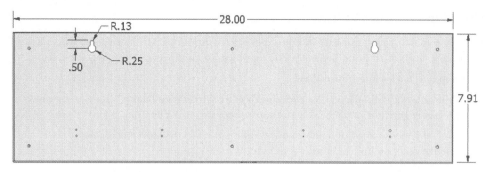

You can print the sheet with the **File | Print** command.

> **TIP** *Inventor provides several AutoCAD-like object snap modes, which are available only during dimensioning. To toggle them, right-click the drawing area, and then choose **Snap Settings** from the shortcut menu.*

Tutorial: Changing AutoCAD Layer Properties

You can change the properties of layers, which has the effect of changing the properties of the AutoCAD objects. To simply toggle layers, follow these steps:

1. In the toolbar, click the **Layer** control. Notice that AutoCAD's layer names are listed near the bottom, following the separator line (------). All the other names above the separator line are Inventor's default layers.

2. To turn off a layer, click the lightbulb icon next to a layer name.

To change other layer properties, follow these steps:

1. Click the **Layer** icon next to the droplist. Notice the Style And Standard Editor dialog box. This dialog box does not, unfortunately, segregate AutoCAD layers, so you will have to hunt them down by name.

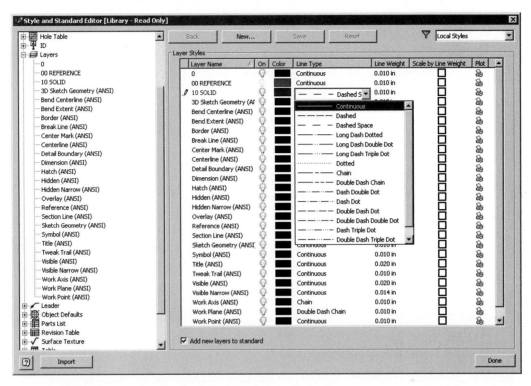

2. In this dialog box, you can change the following properties:

   - Layer name.

   - On-off status.

   - Color.

   - Linetype.

   - Lineweight and lineweight scaling.

   - Plot toggle.

   - New layers (go to the end of the list, and then click **Click Here to Add**.).

3. When done changing properties, click **Done**, and then agree (click **Yes**) when Inventor asks if the edits should be changed. Notice the change to the properties of AutoCAD objects.

You cannot change any other properties or perform any object editing of the AutoCAD drawing.

Tutorial: Reopening Modified Drawings in AutoCAD

The AutoCAD drawing now has annotations added and a view inserted from another Inventor drawing. With the drawing thusly marked up, you can bring it back to AutoCAD. To see what happens, save the drawing before reopening it in AutoCAD. Follow these steps:

1. From the **File** menu, choose **Save Copy As**.

2. In the Save Copy As dialog box, select "AutoCAD Drawings (*.dwg)" from the **Save As Type** droplist.

   In the **File Name** field, change the name to "75 Case From Inventor.dwg" to help identify it. (If Inventor displays the DWG/DXF File Export Options dialog box, click **Finish**.)

3. Click **Save**. Inventor saves the drawing in AutoCAD 2004 format.

4. Switch to AutoCAD, and then open the drawing. Notice that it looks the same as that saved by Inventor.

The results of round-tripping this drawing are as follows:

- The model that was originally designed in AutoCAD was untouched by Inventor; it is fully editable in AutoCAD.

- The Inventor 3D solid model inserted as a base view is converted into a collection of 2D lines and arcs, and is shown in wireframe.

- Annotation objects added in Inventor, such as dimensions and text, are fully editable dimensions and mtext blocks. (Notice the properties of the selected Inventor dimension.) If tables were inserted, they'd be lines and mtext, however.

- The names of Inventor's layers, text styles, dimensions styles, and linetypes are added to the drawing, as can be sees by checking their controls (droplists).

- Model and layout tabs show the same view.

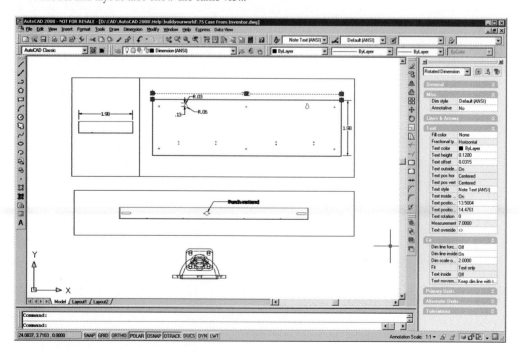

Tutorial: Back Again to Inventor

There's one more test to make: what happens when you modify the Inventor objects in AutoCAD, and then bring the drawing back into Inventor? Let's find out:

1. Change the text of a dimension to read "Modified dimension." Change the linetype and color of the inserted Inventor part.

2. Save the drawing with the **QSave** command.

3. Switch to Inventor, and open the "75 Case From Inventor.dwg" drawing. (If necessary, first close the drawing, and then open it again.) Notice that the Inventor objects now are as uneditable as are AutoCAD objects.

Alternative Endings

While in Inventor, the drawing was saved in AutoCAD's DWG format. But it can also be saved in Inventor's DWG and IDW (Inventor DraWing) formats. Which of these alternative endings to choose?

- **AutoCAD DWG** saves Inventor and AutoCAD objects similarly in both model and layout tabs. Objects can be mildly edited in both CAD packages. Thus, this format is suitable for round-tripping drawings.

- **Inventor DWG** saves Inventor objects in layout tab; while model space contains just the AutoCAD objects. Block-views retain their shading. Other than that, everything is the same as saving in AutoCAD DWG format. See figure below.

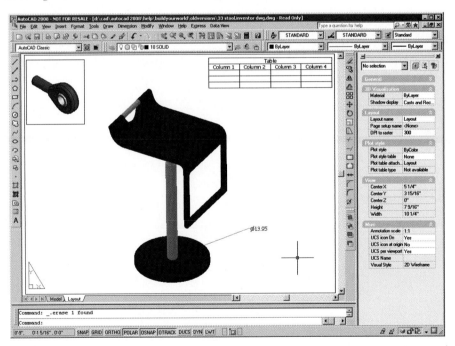

- **Inventor IDW** cannot save non-Inventor objects, and erases AutoCAD objects. After using the Save As command, the drawing is displayed minus any AutoCAD objects. Further, AutoCAD cannot open *.idw* files. Thus, this format is a dead-end for round-tripping.

  Caution: IDW cannot contain any AutoCAD data.

The tutorial showed how Inventor and AutoCAD work with *.dwg* files opened for viewing:

- AutoCAD objects are not converted.

- AutoCAD layers are added, and their properties can be modified.

- Views and annotations can be added by Inventor.

- The changed drawing can be reopened in AutoCAD, edited some more, and subsequently reopened in Inventor.

- Inventor objects can be edited to a limited extent; Inventor styles are preserved.

- Drawings can be saved in AutoCAD DWG or Inventor DWG.

> **TIPS** *If you run Inventor on Microsoft's Vista operating system, be sure to install Inventor 2008 Service Pack 1, because it cures many bugs related to working with .dwg files. For detailed information, read images.autodesk.com/ adsk/files/inventor2008_sp1.txt. Download the 31MB service pack from images.autodesk.com/adsk/files/inventor2008_sp1.msp.*
>
> *For other bug fixes, be sure to monitor the Autodesk Inventor section of support.autodesk.com.*

Converting AutoCAD Drawings to Inventor Format

In the last tutorial, you saw how you can open AutoCAD drawings in "read-only" mode — although you can add as much AutoCAD data as you want. If you want to edit AutoCAD drawings, then you need to *import* them, a process that converts AutoCAD objects into Inventor objects.

In this tutorial, you get Inventor to **convert** another AutoCAD sample drawing. This means that Inventor will translate AutoCAD objects into Inventor objects.

1. Start Inventor, and then from the **File** menu, choose **Open**.

2. As before, change **Files of Type** to "AutoCAD Drawing (*.dwg)" and change the **Look In** droplist to AutoCAD 2008's *help**buildyourworld* folder.

   Choose the "63 Bowsight.dwg" file, but don't click Open just yet.

   > **TIP** *If you cannot find a .dwg file, that's because Inventor may have copied the original to a folder named \Old Versions that it creates for back up. Look for it there under a name similar to "filename 0000.dwg." This naming nicely preserves the original .dwg file, keeping it free of changes by Inventor.*

3. Click **Options**. Notice that the dialog box gives you two choices: open or import.

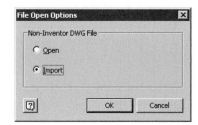

   - **Open** views the drawing in read-only mode, as described in the previous tutorial.
   - **Import** converts the drawing to Inventor format, the topic of this tutorial.

4. Choose **Import**, and then click **OK**.

   Click **Open** to begin importing the drawing. Notice the DWG/DXF File Wizard.

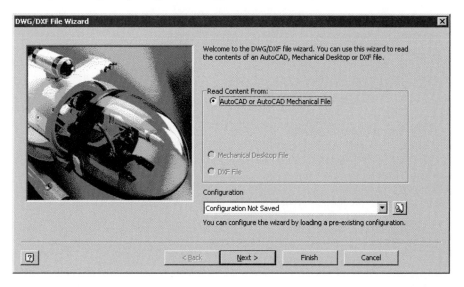

5. To make sure this introductory tutorial works correctly:

    a. Click **Next**.

    b. Ensure **All** and **Model** tab are selected, and then click **Next**.

    c. Under Destination for 2D data, choose **New Part**, and then click **Finish**.

    You learn about this wizard in detail in Chapter 9. Notice that the drawing opens in Inventor's Part environment. (The black dots indicate the centers of arcs and circles.)

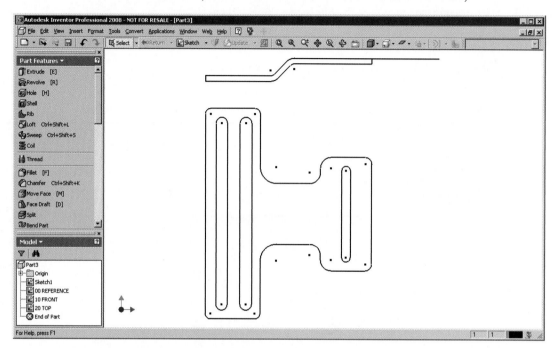

(If you were to open an AutoCAD 3D drawing, it might end up in Inventor looking strange, such as in the figure below. You may be wondering, "Where did those lines come from?" It sure doesn't look like the 3D drawing in AutoCAD. What happens is that Inventor extracts lines and circles that helped define the model.

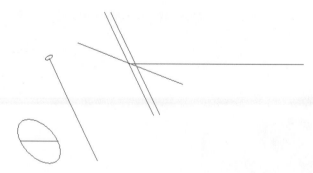

It is important to understand when you import AutoCAD drawings into Inventor, the final result depends on the settings in the wizard. If things look weird, consult Chapter 9.)

The AutoCAD drawing is fully translated into Inventor objects. You can now proceed to edit the parts, features, and sketches as normal.

> **TIP** *There is a second way to import AutoCAD drawings into Inventor. In the Part environment, choose the **Insert AutoCAD Drawing** button in the panel of the Sketch environment.*

No Round-tripping

You cannot save this drawing in DWG format for AutoCAD. The Part environment saves only as an *.ipt* (Inventor ParT) file, and the assembly environment (3D model) only as an *.iam* (Inventor AsseMbly) file.

Try it: both the Save As and Save Copy As commands no longer list .dwg as an available file format.

Round-tripping is possible only if you use Inventor's **Save Copy As** command to save the 3D drawing in SAT format, and then import the *.sat* file into AutoCAD with the **AcisIn** command.

Summary

This chapter was an overview of how AutoCAD and Inventor drawings work together. In the next chapter, you learn in greater detail how to open AutoCAD drawings for viewing in Inventor.

Chapter 8

Viewing AutoCAD Drawings with Inventor

The previous chapter provided overviews of viewing and importing AutoCAD drawings with Inventor. This chapter delves more deeply into what happens when Inventor opens *.dwg* drawings for viewing, and explores the benefits and pitfalls of the process. For example, AutoCAD objects behave differently in Inventor 3D part files than in Inventor 2D drawings, and differently again in Inventor assemblies.

Autodesk provides some information on how AutoCAD drawings act in Inventor. In Help, search for "use AutoCAD geometry," and then click on the Concepts tab.

Here is a synopsis of the help text:

> Inventor directly opens *.dwg* files created by AutoCAD, but it is limited to viewing[1], measuring[2], and plotting[3] objects. AutoCAD objects remain read-only AutoCAD objects[4]. This means that you can view *.dwg* files in Inventor without requiring AutoCAD[5]. Inventor can import AutoCAD objects through the Clipboard[6]: copy the objects in AutoCAD using Ctrl+A followed by Ctrl+C, and then paste them with Ctrl+V in Inventor.

continued

IN THIS CHAPTER

- Viewing AutoCAD objects in Inventor.
- Working with blocks, layers, viewports, text and dimension styles, views, and annotations.
- Making data coexist.
- Understanding problem objects.
- Reusing AutoCAD drawings and blocks.
- Using the Clipboard in AutoCAD and Inventor.
- Round-tripping between AutoCAD and Inventor.

Blocks[7], layers[8], and layouts[9] can be edited in both AutoCAD and Inventor. Text[10] and dimension styles[11] are synchronized: changes made in one are applied to the other.

AutoCAD's layouts are displayed as "sheets" in Inventor[12]. You can create views[13] and place annotations[14] in Inventor; the added data coexists with the AutoCAD data[15].

Viewing, Measuring, and Using the Clipboard

So that's Autodesk's take on the matter. I'm going to expand upon Autodesk's documentation. The phrases I numbered above are referenced below.

When Inventor opens a *.dwg* file created by AutoCAD, it acts like this:

1. **Viewing** the drawing works to a limited extent. Inventor zooms, pans, and rotates 3D models. Inventor opens 3D drawings with the same viewpoint as when they were last saved in AutoCAD. (The Rotate command is found on the panel, but is missing from the View menu. Also unavailable is the useful Look At command.)

   AutoCAD drawings are displayed by Inventor in wireframe only, because Autodesk found that most users are interested only in transferring data from engineering drawings. Shading and hidden-line removal are unavailable; drawings saved with a visual style in AutoCAD are shown in wireframe in Inventor.

   Layouts are displayed as sheets (more later). Xrefs, embedded OLE objects, and attached images are displayed, provided Inventor can find them. DWF files, and DGN files are not displayed, and so cannot be viewed.

2. **Measuring** works with distances and angles. Inventor recognizes objects sufficiently to allow you to select them.

3. **Plotting** is available, and operates as expected. You can print the current sheet, all sheets, or specific sheets; Model mode is considered a "sheet" for printing purposes.

   When Inventor exports 3D drawings in DWF format, they are in 2D only. The drawings can be saved in raster formats (TIFF, JPEG, and so on) with the Save Copy As command.

   When exported as *.dwf* files from Inventor, 3D AutoCAD drawings continue to be seen by DWG Viewer and Design Review as 2D. You cannot rotate the models or make them display in a shading mode other than wireframe.

   (It might appear that the Save command can save AutoCAD drawings in Inventor's *.idw* format, but only Inventor-specific objects are saved; those from AutoCAD are discarded. To save both AutoCAD and Inventor objects, use Inventor's *.dwg* format.)

4. **Non-editable** is the way most AutoCAD objects remain. The only time you can touch them is while measuring them. The exceptions are block positions, layer properties, some layouts, and text and dimension style properties, as described below.

   However, all 2D geometry and some 3D geometry is available for copying, with the exception of viewports and objects seen in viewports.

5. Autodesk notes, "This means that you can view *.dwg* files in Inventor without requiring AutoCAD," but to the limited extent described in note 1, above.

6. Autodesk mentions placing AutoCAD objects through the **Clipboard**. This allows you to place geometry in Inventor for part creation; other AutoCAD concepts are not supported, such as viewports. AutoCAD objects, viewports, and so on, cannot be copied and pasted within Inventor, such as from Model view to Sheet view, or even within the Model or sheet environments. In these cases, the Paste command is missing.

Blocks, Layers, and Viewports

According to Autodesk, the following *can* be edited in Inventor:

7. **Block instances** can be edited in a limited fashion: only the scale, rotation angle, and position can be changed, and then only on sheets. Their content cannot be edited, and blocks in Model mode cannot be accessed. This means that you should copy blocks to a layout tab in AutoCAD before opening the drawing in Inventor. Use the Insert AutoCAD Block command to place instances of blocks in drawings.

    Dynamic blocks are displayed in their current incarnation only; they lose their alternate personalities. Inventor blocks cannot be nested.

    To edit blocks, right-click the block in a sheet (or its name in the Model panel), and then choose **Edit AutoCAD Block**. The AutoCAD Blocks dialog box appears with the options for changing scale (size) and rotation angle.

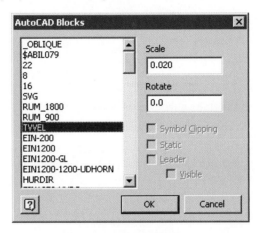

To move a block, grab it by its insertion point (the green dot, as shown below), and then drag it to its new location.

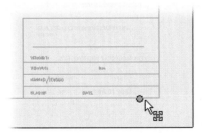

Although it appears that the block's color and lineweight can be changed through Inventor's Properties dialog box, this is not the case; changes you make are ignored. A limited workaround is to use the Layer dialog box to change color and so on, but this works only when objects in the blocks have ByLayer properties. Inventor does not have the ByBlock property.

> **TIP** *If you need to edit the content of blocks, use this workaround: right-click the block, and then choose **Edit in AutoCAD**. The drawback is that the drawing must be closed in Inventor so that the entire drawing is opened in AutoCAD — not just the block. Once done in AutoCAD, save and close the drawing, and then open it again in Inventor.*

To view block names, open the Model panel's trees for model or sheets, and then open **AutoCAD Blocks**.

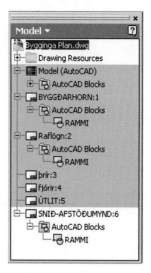

(The gray area in the figure above indicates non-activated sheets.)

8. **Layer** properties can be changed, although Inventor layers don't have as many properties as AutoCAD.

To view layer names and change their properties, use the Standard toolbar, which operates like the Layer control in AutoCAD.

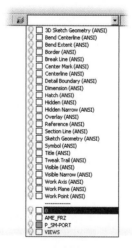

Notice that the list begins with the names of Inventor's standard-based layers, and then lists AutoCAD's at the bottom, because they are not part of the standard. The ☼ light bulb icon toggles visibility, while the ☐ square indicates color.

To modify color and other properties, click the **Layer** button. Unlike the droplist, the dialog box does not segregate AutoCAD layer names, which is unfortunate.

9. **Layouts** are displayed as "sheets" in Inventor. Sheets are created automatically when the AutoCAD drawing contains layouts. Once in a while, however, I found that an AutoCAD drawing would display blank sheets for its layouts. (The technical editor found that using AutoCAD's Recover command helps layouts display correctly in Inventor.)

Sheet names are copied from AutoCAD's layout names, such as "ISOA1FinalPlot:1" and "Utility:2." The number suffixes (:1 and :2) indicate the sheet number, which is needed for plotting. Sheets are listed in the Model panel by name, as illustrated by the figure below.

Multiple AutoCAD paper space viewports are displayed correctly by the sheets. Model space viewports are ignored.

You can add views of other Inventor models to sheets, but not of AutoCAD, because the Base View command does not recognize AutoCAD drawings.

TIP *To switch from Model to a named sheet (layout) mode, right-click the name of a sheet, and then choose* **Activate**.

Text & Dimension Styles

Autodesk says that text and dimension styles are synchronized: changes made in one are applied to the other.

10. **Text styles** are displayed correctly, even though Inventor does not handle all of AutoCAD's style properties. Obliqued and vertical text, for instance, are not available in Inventor, yet are displayed correctly.

    Special characters, such as %%c (diameter symbol), are displayed correctly. The codes for overlining and underlining are ignored, however.

    Fonts are displayed correctly, even SHX fonts that Inventor does not normally support; Inventor substitutes TTF (TrueType) equivalents for SHX files.

    To view text styles, choose **Style and Standards Editor** from the **Format** menu. In the tree, open **Text**.

11. **Dimension styles** are displayed correctly, as are tolerances and leaders. To view dimension styles, choose **Style and Standards Editor** from the **Format** menu. In the tree, open **Dimension**. As with layers, text and dimension styles from AutoCAD are not singled out.

Views & Annotations

Autodesk says that views and annotation can be added to drawings, with the added data coexisting in both CAD programs.

12. **Views** (viewports) can be created in sheets of Inventor assemblies, parts, and presentations.

    To create views, use the **Base View** command, and then choose an Inventor file from the Open dialog box.

13. **Annotations** can be added in the Drawing environment. Annotations include dimensions, text, leaders, and tables.

The panel makes it appear as if other commands work with AutoCAD objects, but don't. This includes Balloon, Parts List, and Retrieve Dimensions. These would work only with mixed AutoCAD-Inventor drawings.

Text and other annotations from AutoCAD cannot be edited. AutoCAD 2008's new annotative text is like dynamic blocks: only the version in effect is displayed by Inventor.

To add annotations, click the **Drawing Views Panel** droplist, and then choose "Drawing Annotation Panel."

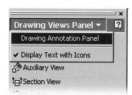

Coexisting Data

14. Autodesk notes that some of "the added data coexists with the AutoCAD data." This means that you can save the modified drawing in Inventor's *.dwg* format, and then open it in AutoCAD. The word "Some" refers to the critical associative view data (information required to support associative updates) when models are modified. This is unavailable through traditional export and translator techniques.

When you edit the Inventor data (blocks, annotations) in AutoCAD, and then open the drawing in Inventor again, *Inventor does not display edits* made to the view-blocks (such as with BEdit). The view-blocks are displayed as they were before being edited in AutoCAD. This is not unreasonable, because the intent is that Inventor content in AutoCAD files should still be associated to Inventor part files. Thus, any editing of Inventor blocks should be performed only in Inventor.

For example, you have a very large machine currently documented in 2D AutoCAD. You want to revise parts of it using Inventor, but don't want to model the entire machine in 3D. No problem: create the new parts in Inventor, delete the 2D objects that represent them in the AutoCAD, and then insert Inventor views. The new views remain associative back to the Inventor model files.

While in AutoCAD, you can expect the following from the drawing:

• The views and annotations are stored as blocks at the same scale as in Inventor.

• Shaded views retain their shading.

• All of Inventor's layer names and linetypes, as well as dimension and text styles, are named in the drawing.

• Annotations must be exploded before editing them as mtext.

• Dimension are non-associative.

• Blocks must be opened with the BEdit command for editing in the Block Editor (see figure below); they are not accessible in any other way.

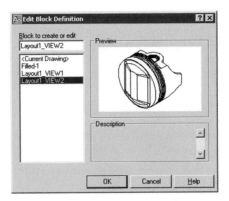

You'll see a dialog box that reports: "Editing drawing view geometry outside of Inventor is not supported. The edited views will be restored to their original state."

AutoCAD edits made to annotation, however, are preserved and displayed in Inventor. But they can no longer be edited in Inventor.

TIPS *AutoCAD recognizes Inventor's lock file. When drawings are open in Inventor, they can be opened only in read-only mode in AutoCAD. The reverse, however, is not true, because the AutoCAD drawing is already in read-only mode in Inventor.*

*Inventor's **File | Open** command distinguishes between its own .dwg files and those of AutoCAD. If you cannot see the drawing file you are looking for, change the Files of Type to "Autodesk Inventor Drawings (*.dwg, *.idw)" or "AutoCAD Drawings (*.dwg)."*

You will see objects in the preview image displayed by Inventor's Open dialog box, even if they are not displayed by Inventor, such as shading. That's because the preview image is a raster image generated by AutoCAD; it has no relationship to Inventor's ability to display or import objects created by other software programs.

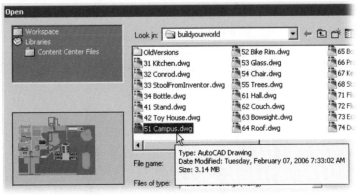

How Inventor Handles AutoCAD Objects

I continued to test Inventor by opening a number of other 3D and 2D drawings, including some used for architecture and terrain mapping. Inventor displayed nearly everything correctly, but I did find some problems.

The *solar.dwg* drawing illustrates the ability of AutoCAD to draw the Solar System full size. This drawing shows that Inventor correctly handles drawing with very large units. The famous *teapot.dwg* drawing is made completely of 3D faces: Inventor correctly handles them.

All of AutoCAD's 16.7 million colors are supported, as the *colorwh.dwg* sample drawing shows. Plot styles are not supported, as shown (or rather, not shown) by the *Plot Screening and Fill Patterns.dwg* sample drawing.

Although Inventor does not support SHX fonts, it correctly displays text made with them, as proven by the *tablet.dwg* sample drawing. Field text is displayed correctly, but it become regular text and is not updatable.

For a definitive list of supported objects, see the end of this chapter.

Problem Objects

Vertical lines in old 3D drawings were missing. I think that the vertical lines are made of points that have been given thickness. This was a method of simulating 3D lines in older releases of AutoCAD, and is something Inventor perhaps doesn't understand.

When Inventor reads blocks, it converts attributes to text. Curiously, Inventor has a right-click menu item named "Edit Attributes," but it is grayed-out, even when you select blocks with attributes. Similarly, database links are ignored, as are links between tables and data sources. Tables are correctly displayed.

Inventor cannot open AutoCAD sheet sets, for it does not handle *.dst* files. However, Inventor does open the individual *.dwg* files that make up sheet sets.

Named views and table styles are not recognized by Inventor. DWF and DGN underlays are discarded.

Viewing AutoCAD Blocks

Inventor reports the names of the blocks it finds in AutoCAD drawings. To see this list and view the names of block instances, follow these steps:

1. Open an AutoCAD drawing in Inventor with the **File | Open** command.

   (In the Open dialog box, change **Files of Type** to "AutoCAD Drawings (*.dwg)." Select an AutoCAD drawing file, and then click **Open**.)

2. In the history tree, open **Model (AutoCAD)**, and then open **AutoCAD Blocks**. Notice the list of block names.

TIPS *Block names mostly are located under Model of the history tree. Usually, the only blocks to appear under Layout are those related to paper space, such as drawing borders and title blocks.*

*Inventor initially lists block names in the order that they appear in the drawing. To sort the list alphabetically, right-click **AutoCAD Blocks**, and then select **Sort by Name** from the shortcut menu.*

*You can see a list of every block definition in the file (and insert new instances) by looking under **AutoCAD Blocks** in the browser's Drawing Resources folder.*

3. Select the name of a block. Notice that the block is highlighted in the drawing.

4. If the drawing is cluttered, you might not see the block being highlighted. To get a better view of it, right-click the block's name. From the shortcut menu, select **Zoom To**.

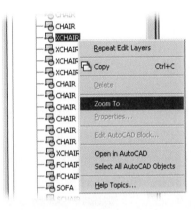

Notice that Inventor zooms into the block so that it fills the viewport.

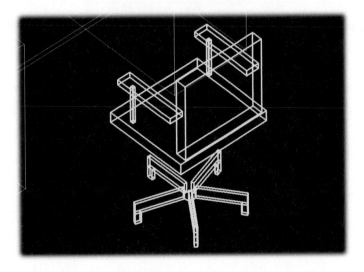

Layers Imported from AutoCAD

To see the names of AutoCAD layers, click the **Layers** droplist. Notice that there are two sets of layer names: the top portion contains Inventor's layer names found in the current standard, while the lower portion lists the nonstandard layer names, which include those imported from AutoCAD. In the figure at right, AutoCAD's layers begin with "0" below the dashed line.

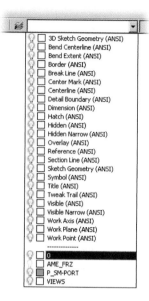

To turn off layers, click the ♀ light bulb icon.

To change the color of layers, follow these steps:

1. Click the ⬚ **Layers** icon next to the layer droplist. Notice the dialog box (illustrated below).

2. Unfortunately, AutoCAD layers are not segregated from Inventor's, so you'll have to hunt down the one(s) you need.

3. When you find one you want, click the color sample to open the Color dialog box.

4. Choose a color, and then click **OK**.

This dialog box also allows you to change the visibility, linetype, lineweight, and plot status of imported AutoCAD layers.

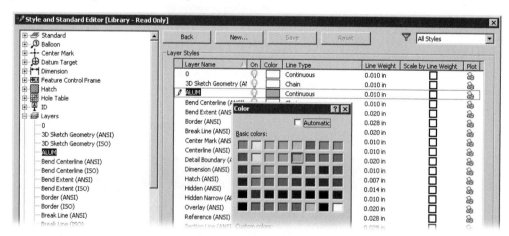

5. When done, click the **Save** button, and then exit the dialog box. The AutoCAD drawing takes on the new look defined by the changes to its layers.

Viewports and Externally-Referenced Drawings

Inventor handles drawings with multiple viewports differently, depending where the viewports are located. In summary, the following happens:

- When multiple (tiled) viewports exist in **model** space. Inventor discards them, and displays just the model in a "single" viewports.

- When multiple viewports exist in **layout** mode, then Inventor displays all of the viewport in its Drawing environment. In the figure below, I highlighted the viewports by painting them in white.

Xrefs are also displayed correctly. Inventor opens them in the Drawing environment, with each xref shown in its own base view located correctly on a sheet.

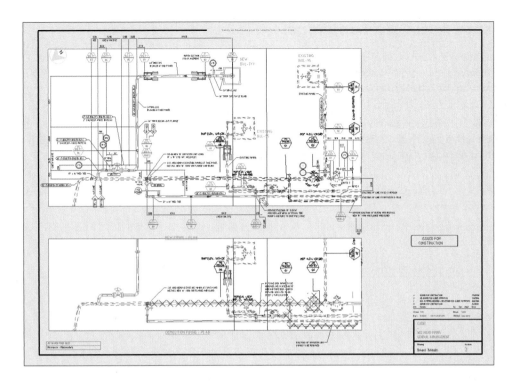

Measuring AutoCAD Drawings

You can make linear and angular measurements of AutoCAD objects. Inventor's measurement tools are intelligent, and work like this:

1. From the Drawing Review panel, select **Measure Distance**. Notice that the Measure Distance dialog box appears, as illustrated below.

2. Move the cursor over lines in the drawing. As you do, Inventor highlights the feature in red. If it is a curve, the radius or diameter lines are also shown.

3. Click the cursor on the object. Notice that Inventor reports the measurement in the dialog box.

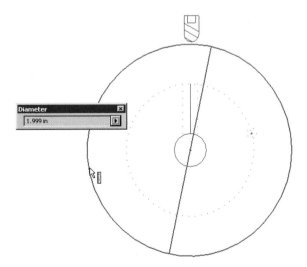

Inventor also measures angles between two points. See Chapter 5 for a detailed tutorial.

Reusing AutoCAD Drawings

AutoCAD drawings can be used as templates for new Inventor drawings. Alternatively, with an AutoCAD drawing placed in Inventor, you can draw over it. Inventor sufficiently recognizes individual AutoCAD objects so that you can use object snaps to draw accurately.

(This reminds me of one way in which raster images are handled. Instead of converting the raster dots to vector lines, the images are opened in AutoCAD, and then the operator erases unwanted portions and draws over additions to modify the "drawing." The resulting hybrid raster-vector image is saved and printed. This approach is reasonably good for such situations as showing new additions to existing machinery.)

If you have created *.dwt* template files and *.dwg* block libraries for AutoCAD, you can reuse them in Inventor. This lets you carry over presets. The ones that Autodesk mentions in its documentation are blocks, layers, text styles, and dimension styles.

I tested Inventor to see if anything else copied over. I checked this fuller list of AutoCAD named objects and properties:

AutoCAD Named Objects	Transferred to Inventor
Blocks	Use directly or convert to symbols; attributes converted to text.
Color books	No: Inventor lacks color books.
Dimension styles	Names and properties.
Gradients	No: Inventor lacks gradients.
Hatch patterns	No: patterns hardcoded in Inventor.
Layouts	Names and some properties, such as sheet size.
Layers	Names and properties.
Linetypes	Yes: when objects are imported or pasted, the linetypes are copied, both simple and complex; Inventor reads *.lin* files.
Lineweights	No: hardcoded in AutoCAD and Inventor.
Materials	No.
Plot styles	No: Inventor doesn't use plot styles.
Shapes	No.
Text styles	Names and properties.
Multiline styles	No: Inventor lacks multilines.
Table styles	No.
Viewports	Paper space only; names, positions, and scales.
Visual styles	No: visual styles hardcoded in Inventor.

While layer and style names and properties are added automatically, other items require hand tweaking. These items need to be *imported* from the *.dwg* files, not just opened:

Borders — select "Border" in the DWG import wizard.

Blocks — select "Symbol" in the DWG import wizard; just one block can be imported at a time; i.e., the entire drawing is treated as a block.

Title blocks — select "Title Block" in the DWG import wizard.

See Chapter 9 on how to import AutoCAD drawings, and convert their objects to Inventor format.

The following tutorials show you how to reuse AutoCAD drawings for templates and base plans.

Tutorial: AutoCAD Drawings as Templates for Inventor

Creating Inventor templates from AutoCAD templates takes three stages: (1) opening the drawing in Inventor, (2) specifying object defaults with the Styles Editor, and (3) saving the drawing in Inventor's *template* folder.

> **TIP** AutoCAD keeps its template files in an inconveniently hidden folder. Because this is a hidden folder, you won't see it in Windows Explorer or file dialog boxes until you first turn on the **Show Hidden Files and Folders** option in Windows Explorer's **Tools | Folder Options | View** dialog box.
>
> By Microsoft dictate, software vendors must place application files unique to individual users in hidden folders; it's not clear why the files are kept hidden from owner-users.

1. AutoCAD template files use the *.dwt* extension, one that Inventor does not recognize. Before bringing template drawings into Inventor, you have to change their extension to *.dwg*.

    Rather than renaming the files, I recommend making copies, and then renaming the copies. In Windows Explorer, you can make copies of files like this:

    a. In Windows Explorer, navigate to the *C:\Documents and Settings*<logon name>*\Local Settings\Application Data\Autodesk\AutoCAD 2008\R17.1\enu\Template* folder, and then choose a file name, such as "Tutorial-mMfg.dwt."

    b. Hold down the **Ctrl** key.

    c. Drag the file name to the end of the list of names. Notice that the cursor has a + icon; this means Windows will make a copy of the file.

    d. Let go of the **Ctrl** key. Notice that Windows makes a copy, naming it "Copy of Tutorial-mMfg.dwt."

    e. Right-click the file name, and then choose **Rename** from the shortcut menu.

    f. Rename the file to "Tutorial-mMfg.dwg."

    Repeat this procedure for every *.dwt* file you want to use as a template in Inventor.

2. In Inventor, choose **Open** from the **File** menu.

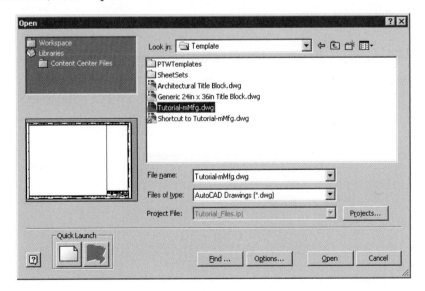

- In the **File of Type** droplist, choose "AutoCAD Drawing (*.dwg)."

- From the **Look In** droplist, go to the folder containing the AutoCAD drawing to be opened.

- Select the *Tutorial-mMfg.dwg.* file to use as a template, and then click **Open**.

Notice that the drawing opens in Inventor's Drawing environment. In addition, Inventor imports all AutoCAD layer names and properties, as well as dimension and text styles. For example, you can click the layer droplist to confirm the names of AutoCAD layers at the end of the list.

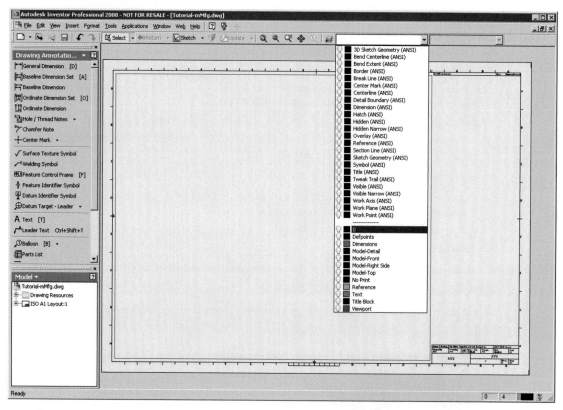

3. Inventor normally includes all style settings, unlike AutoCAD. This means you may want to narrow down the styles in this template drawing. To do so, choose **Styles And Standard Editor** from the **Format** menu.

4. In the dialog box, choose **Default Standard (ANSI)**, and then click **New**.

5. In the New Style Name dialog box, give the style its name, such as "MMF Mech," and then click **OK**.

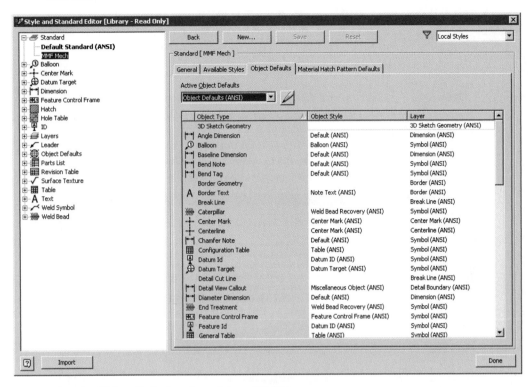

6. Select the **Object Defaults** tab, and then click **New**.

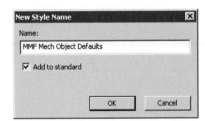

•   Give the style a name, such as "MMF Mech Object Defaults."

•   Turn on the **Add To Standard** option, which ensures it is used by the new standard.

•   Click **OK**.

7. You can now edit the new style. For example, use the **Available Styles** tab to specify the layers, dimension and text styles, and other styles available. Clear the check boxes of styles you don't want used.

8. Click **Save**, and then **Done** to save the style changes.

9. From the **Format** menu, choose **Purge Styles** to remove unused standards and styles from the template.

10. From the **File** menu, choose **Save Copy As** to save the template file into the template storage folder.

- In the **Save In** droplist, select the appropriate folder for templates, probably *C:\program files\autodesk\inventor 2008\templates.*

- Name the template something significant, such as "MMF mfg Template.dwg."

- Click **Save**.

The new template is now available for creating new drawings.

> **TIPS** *Unlike AutoCAD, Inventor does not have a file extension that identifies template files; instead, templates are identified by being in a folder named C:\program files\autodesk\inventor 2008\templates. The path is defined by:*
> - *From the **Tools** menu, choose **Application Options**, and then **Files** tab.*
> - *Or, in the **Files** menu, choose **Projects**, and then expand **Folder Options**. Right-click on the **Templates** item, then choose **Edit** from the context menu.*

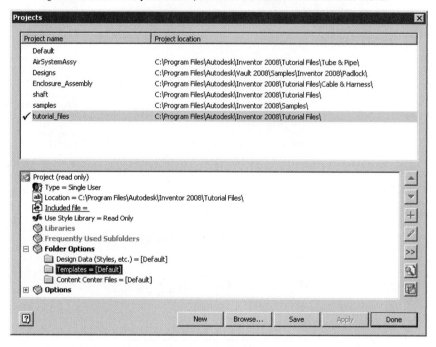

> *Use either method to relocate the template folder to a network drive.*
>
> *You can add template tabs to the New File dialog box by creating new folders in the Template folder with Windows Explorer. This lets you create templates for different projects and clients, with each set of templates in their own tab.*

Reusing AutoCAD Blocks in Inventor

AutoCAD blocks can be reused in Inventor drawings. Unfortunately, Inventor cannot import block libraries — .*dwg* files filled with blocks, such as those found in AutoCAD's *sample**designcenter* and *sample**dynamic blocks* folders. (Inventor calls blocks "symbols.")

Blocks must be imported into Inventor; if you open them, they are unusable. (Recall that Inventor can either open .*dwg* files for viewing, or import them for conversion.) Here's what happens when blocks are imported:

- Block definitions are converted to symbols of the same name.
- Block instances are inserted in the Sketch environment as 2D sketches.
- Nested blocks are exploded and imported as a single block.

Follow these steps to import blocks from AutoCAD:

1. Before bringing your block library files into Inventor, you need to save each block in its own .*dwg* file. Open the drawing in AutoCAD, and then use the **WBlock** command to export each block to its own drawing file. I recommend that you create subfolders to hold the dozens of resulting drawings.

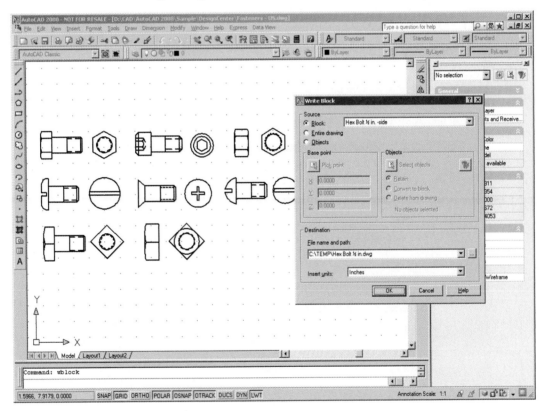

2. Switch over to Inventor, and then select **Open** from the **File** menu.
3. In the Open dialog box, do the following:
   a. Change the **Files Of Type** droplist to "AutoCAD DrAWings (*.dwg)."
   b. Use the **Look In** droplist to go to the folder holding the block-drawings.
   c. Select a .*dwg* file.

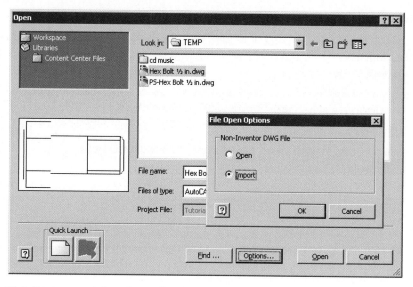

d. Click **Options**, and then choose **Import**.

e. Click **OK**, and then click **Open** to exit the two dialog boxes.

(Ignore the dialog box warning about projects by clicking **OK**.)

4. In the DWG/DXF File wizard, click **Next**, and then click **Next** again.

5. In the Import Destination Options dialog box, ensure that **Symbol** is selected under Destination for 2D Data. If this option is not turned on, then the block is exploded and turned into individual lines, arcs, and circles.)

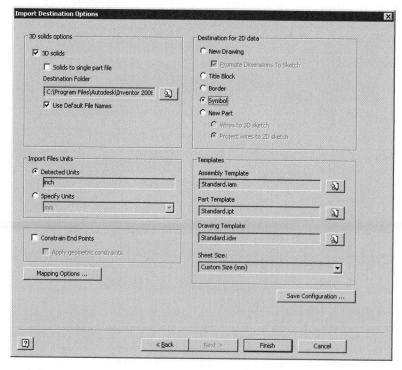

(Chapter 9 discusses this dialog box in detail.)

6. Click **Finish**. Notice that the block appears in Inventor's Drawing environment.

    You'll find the block's name listed under Sheet:1 | Sketched Symbols of the Model panel.
    (It's also listed under Drawing Resources | Sketched Symbols.)

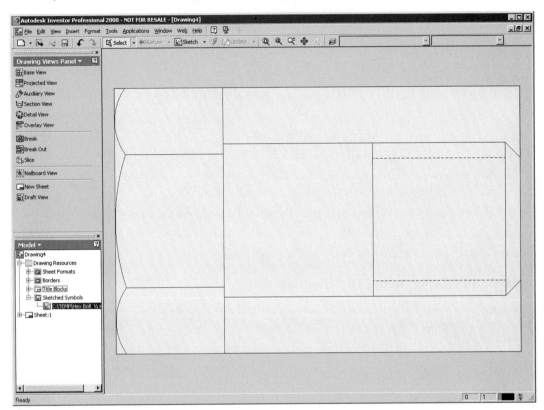

Now that the block-symbol library is imported into Inventor, you can do the following tasks:

-   Right-click a symbol name under Sheet:1 | Sketched Symbols, and then choose **Edit Definition**. The symbol opens in the equivalent of AutoCAD's Block Editor. Notice all of the 2D sketching and editing commands available in the panel.

-   While editing the definition, select one or more objects, and then right-click. From the shortcut menu, choose **Properties**. You can specify the color, linetype, linetype scale, and lineweight for the symbol.

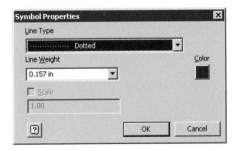

    When you are done editing the symbol definition, click the **Return** button on the toolbar.

-   Right-click its name under **Sheet:1 | Sketched Symbols** again, and this time choose **Edit**

Symbols. In the Symbols dialog box, you can change its scale, rotation angle, and other properties that affect its insertion.

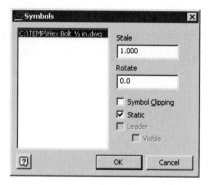

- You can copy and paste symbols between drawings: right-click a symbol, and then choose **Copy**. Switch to another drawing, and then right-click and choose **Paste**.

- Use the **File | Save As** command to save the symbol as an Inventor drawing file (*.idw*).

To make the sketch-blocks available in all new drawings, add them to template files.

Tutorial: Annotating AutoCAD Drawings

A task Inventor is well suited to is annotation. While AutoCAD 2008 added numerous annotation-oriented features, such as annotative text and multiline leaders, Inventor is still better at annotating mechanical drawings than is AutoCAD.

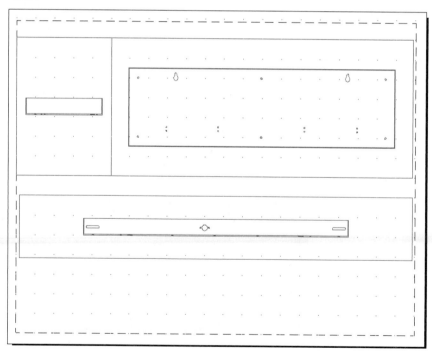

With that in mind, you may want to use Inventor for the sometimes tedious task of adding symbols, dimensions, and so on, to AutoCAD drawings. Here's how to do it:

1. For this tutorial, open the AutoCAD sample drawing *75 Case.dwg* in Inventor. (You'll find this drawing in AutoCAD's *\Sample\Sheet Sets\Manufacturing* folder.) You may need to prepare it in AutoCAD by creating viewports in layout tab, as illustrated above.

   Notice that the drawing is opened in Drawing Views mode.

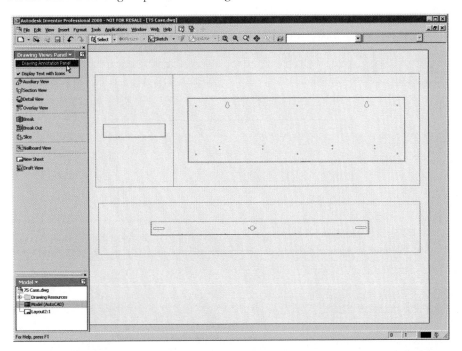

2. Switch to annotation mode: click the **Drawing Views Panel** droplist, and then choose "Drawing Annotation Panel." Notice that all of Inventor's annotation commands become available, such as dimensioning, leaders, tables, and text.

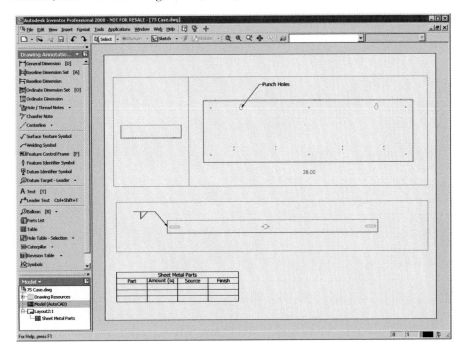

3. Use the **File | Save As** command to save the drawing in "Inventor Drawing (*.***dwg***)" format. (Don't use the similarly named "Inventor Drawing (*.***idw***)" format, because it erases all AutoCAD objects.)

4. Open the drawing in AutoCAD. The annotations added by Inventor are stored as blocks, which can be edited after exploding them.

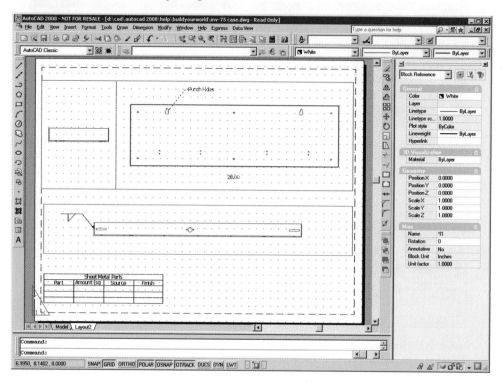

Other Menus and Methods

Inventor provides a variety of ways to access AutoCAD drawings, such as through the Clipboard and through drag and drop.

Using the Clipboard in AutoCAD

The Clipboard is another method for bringing AutoCAD objects to Inventor. Since the Inventor package includes AutoCAD, this method can be useful, even if it requires that AutoCAD and Inventor be running at the same time. Note that pasting into Inventor's 2D Drawing environment file is different from pasting into its 3D Part environment.

Follow these steps:

1. Open a drawing in AutoCAD, and select the objects to be copies. (To select all objects, press **Ctrl+A**).

2. Copy them to the Clipboard by pressing **Ctrl+C**.

3. Switch to Inventor, which should have a drawing open.

4. Press **Ctrl+V** to paste the AutoCAD objects in Inventor.

> **TIP** *As an alternative to Ctrl+V, you can select **Edit | Paste Special** to select the format in which to paste the data.*

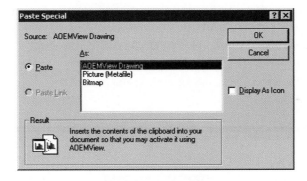

The selections are:

*   **AOEMView Drawing** — *pasted as AutoCAD data; the same as when using Ctrl+V.*
*   **Picture (Metafile)** — *pasted as a block in WMF (Windows Meta File), a vector format.*
*   **Bitmap** — *pasted as a raster image.*

The Clipboard viewer reports that the AutoCAD objects are saved in the following formats: Picture, Bitmap, Enhanced Metafile, DIB Bitmap, DataObject, Embed Source, Native, OwnerLink, Object Descriptor, AutoCAD.R17, and Ole Private Data.

Using the Clipboard in Inventor

Autodesk's online help mentions using the Clipboard with imported AutoCAD drawings, but doesn't provide details on what's possible. To check for myself, I selected the 3D model of the steering wheel in Inventor, right-clicked, and then chose **Copy** from the shortcut menu.

The Windows Clipboard viewer reported that the data was copied in "AutoCAD R17" format only. This means it cannot be pasted into any applications other than Autodesk's; for example, I couldn't paste the image into a word processing document or a paint program.

I attempted to paste the AutoCAD data into several of Inventor's environments to see which ones would work:

Inventor Environment	Paste AutoCAD Data
Parts	Yes; pasted data appears in 2D plan view (top view).
Assemblies	No, because only Inventor parts are allowed.
Drawing	Conditional yes; before pasting, choose **Draft View** from the panel. Solid model is pasted in 2D plan view.
Presentation	No, because only Inventor parts are allowed.

I found I could also paste the data from Inventor back into AutoCAD, but it appeared as a 2D block in plan view. When I exploded the block, it consisted of lines, arcs, and circles. The 3D information was lost.

Drag and Drop

A third method of opening .*dwg* files is to drag them from Windows Explorer into Inventor. When Inventor has no files open, this action opens the drawings in the Drawing Review environment.

When Inventor has drawings open, you can't drag .*dwg* files into existing drawings. Instead, drag the .*dwg* file name onto Inventor's title bar, where it is then opened in Drawing Review.

Inventor files, other than Inventor's .*dwg*, cannot be dragged into AutoCAD.

Shortcut Menu

After the .*dwg* file is opened in Inventor, you can select it, and then right-click for a shortcut menu that lists options not available on the panel:

Copy – copies AutoCAD objects to the Clipboard as 2D objects in plan view.

Open in AutoCAD — closes the drawing in Inventor, launches AutoCAD, and then displays the drawing. Before returning it to Inventor, the drawing must be closed in AutoCAD. I am not sure why the drawing needs to be closed, since it is effectively in read-only mode in Inventor.

Select All AutoCAD Objects — selects objects that came from AutoCAD, ignoring those created by Inventor.

How Inventor DWG Files Act in AutoCAD

The goal of creating *.dwg* files from Inventor is to simplify the publishing of engineering and manufacturing drawings from Inventor. The expectation is that you complete drawings in Inventor (which is faster than drawing them in AutoCAD), and then publish them in DWG format for those who use AutoCAD.

Once Inventor drawings are published as AutoCAD copies, the copies cannot be touched to prevent uncontrolled changes made on the shop floor and elsewhere. This is why Inventor views are exported as blocks, and can be reviewed, measured, and plotted only. (You can also explode and delete the objects created in Inventor.)

Inventor can create and save files in DWG format. The data is saved in a version of DWG that *looks* the same in both Inventor and AutoCAD, but adds Inventor-specific data as custom objects — much in the same way that Mechanical Desktop and AutoCAD Architect do.

Since you cannot edit Inventor's dialect of DWG, Inventor's Open dialog box has two entries for *.dwg* in the **Files of Type** droplist: one for Inventor and one for AutoCAD — as highlighted by the figure below.

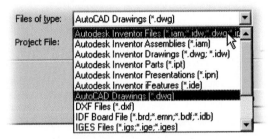

Inventor DWG-based drawings look exactly the same in both programs. But what happens behind the scenes is that Inventor creates block definitions for every drawing view and sketch in the file. (As you edit them in Inventor, the block definitions are updated.)

After the Inventor DWG-format file is opened in AutoCAD, you can use the Insert and ADCenter commands to place the blocks, or use the XRef command to reference them in other drawings. Use the BEdit command to edit the blocks, which curiously enough, is the only way to view the names of view-blocks. As noted earlier, edits do not stick when you go back to Inventor, because the intention is that you use Inventor to edit parts and related the view-block; therefore they are anonymous blocks, like hatches in AutoCAD.

In the figure below, DesignCenter lists preview images of the Inventor-generated blocks, while the Properties palette lists information about the selected block.

> **TIPS** *Only Inventor .dwg drawing files can be exported to AutoCAD. Assembly .iam, part .ipt, and presentation .ipn files cannot be exported to AutoCAD.*
>
> *These limitations are deliberate, and solve the problem of creating and distributing drawings. But, there is a workaround: use Inventor's **File | Save Copy As** command to export the assemblies and parts as SAT files, and then import them into AutoCAD with the **AcisIn** command.*

The following table shows which of Inventor environments can be exported to AutoCAD:

Inventor Environment	File Extension for Inventor	Save Copy As for AutoCAD
Part	IPT	SAT
Assembly	IAM	SAT
Inventor drawing	DWG, IDW	Inventor Drawing (DWG), AutoCAD Drawing (DWG), DXF
Presentation	IPN	-- *none* --
AutoCAD drawing	DWG*	Inventor Drawing (DWG), AutoCAD Drawing (DWG), DXF
Sheetmetal flat patterns	IPT**	SAT
Sketches, planar faces	...	***

Notes:
*) The Save As command can be used only the first time a .dwg template is saved to an .idw file and the first save of an existing .dwg. Once saved, the Save As or Save Copy As commands can be sued to save to the other file type.

***) Right-click the flat pattern in model browser, and then choose **Save Copy As** SAT, DWG, or DXF.*

****) Right-click the sketch in model browser, and **Save Copy As** DWG or DXF. DWF is compatible back to AutoCAD 2000; DXF is compatible back to AutoCAD Release 12.*

The figure below shows an Inventor drawing (**.dwg*) in Inventor's Drawing environment...

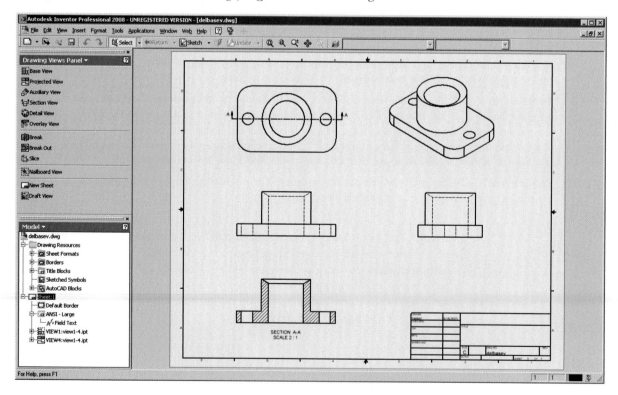

... and the same drawing opened in AutoCAD.

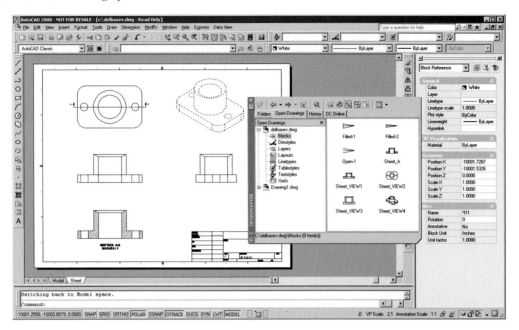

Round-tripping Between AutoCAD and Inventor

You can work with a single *.dwg* file between AutoCAD and Inventor. As you do, they create their own objects and store them as a *.dwg* file. Here is one scenario:

1. Open a DWG-format Inventor drawing in AutoCAD, and then add AutoCAD geometry in model space of a layout tab.

2. Save the file, and then open it Inventor. The AutoCAD-added geometry appears just as it did in AutoCAD.

3. Next, you can add Inventor geometry to the file.

4. Save it, and then open it in AutoCAD.

Thus, the file can go back and forth as needed. There is a limitation, however: only a very few objects and tables can be *edited* in both programs. These include:

- Layers.

- Blocks*.

- Text styles.

- Dimension styles.

- Linetypes.

- Sheet names (known as layouts in AutoCAD).

- Sheet sizes.

> *Note*:
> *) Blocks can be placed in Inventor drawings, but they cannot be defined or edited in Inventor.*

TIPS *While Inventor-generated blocks (symbols) cannot be exploded in AutoCAD, you can use object snaps to aid in drawing accurately with AutoCAD objects in relation to the symbols, as illustrated below.*

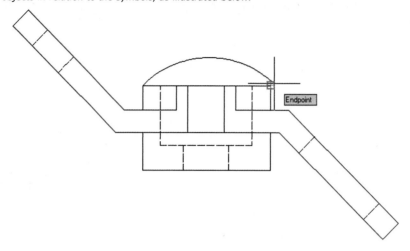

Use the Layer command to change the color and linetype of the blocks. Inventor places solid lines of symbols on the "Visible Edges" layer, while hidden lines are on the "Hidden Edges" layer.

Summary

This chapter described how to open AutoCAD files for viewing in Inventor. You learned some of the ways to exploit Inventor for AutoCAD, and how to round-trip drawings.

Opening drawings is just one way of working with both CAD packages; the next chapter describes methods for importing AutoCAD drawings, in which Inventor converts AutoCAD objects into its own format.

Chapter 9

Importing AutoCAD Drawings into Inventor

Inventor can open AutoCAD drawings for viewing and plotting, as described by the previous chapter. It also imports *.dwg* files, converting AutoCAD objects into Inventor's format, as explained in this chapter.

You might want to import AutoCAD drawings for several reasons:

- Reuse 2D AutoCAD drawings as the basis of sketches in Inventor.
- Reuse 2D title blocks, drawing borders, and other elements originally drawn in AutoCAD.
- Import 3D models created previously in AutoCAD.

(Recapping briefly, *opening* drawings results in perfect images that cannot be edited; *importing* them results in editable drawings that may have lost some data. The reason that two approaches are needed is due to the differences between Inventor and AutoCAD file formats.)

IN THIS CHAPTER

- Importing AutoCAD drawings by two methods.
- Converting AutoCAD files to Inventor.
- Exporting Inventor files to AutoCAD.
- Recovering import problems with Design Doctor.
- Referring to comprehensive Import and export destination references.
- Learning what works, and what doesn't.
- Exporting Inventor environments.

Three Ways to Import AutoCAD Drawings

Inventor provides three ways to import (convert) AutoCAD drawings:

1. **Open | Import** — Choosing the **Import** option of the **Open** command lets Inventor import the 2D and/or 3D components of AutoCAD drawings as sketches, parts, drawings, draft views, and/or assemblies, depending on the options you select.

2. **Insert AutoCAD File** — In the sketch modes of the Part and Drawing environments, use the Insert AutoCAD File command to add editable AutoCAD objects to the sketch. Only 2D objects are imported; 3D objects are ignored.

   This command is easier than Open, because it presets the import wizard's settings, depending on the environment in which you start.

3. **Edit | Paste** — The easiest way to reuse AutoCAD 2D data in Inventor's Part environment is to press **Ctrl+C** in AutoCAD to copy objects to the Clipboard, and then press **Ctrl+V** (Paste) in Inventor. Lines, circles, arcs, and so on become sketch geometry; AutoCAD dimensions become Inventor parametric dimensions.

   This method may be less advantageous for large drawings, because the import wizard lets you choose the layers and objects to bring into Inventor.

   (AutoCAD objects can be pasted into Inventor's Drawing environment, but then they cannot be edited; 3D objects are flatted to 2D.)

Going in the other direction, Inventor has two formats in which to export its drawings to AutoCAD:

1. **AutoCAD DWG** — the Save Copy As command exports Inventor 2D drawings in AutoCAD's flavor of DWG. Inventor objects are stored as blocks; AutoCAD objects added to the Inventor drawing return as AutoCAD objects. All show up in Model and layout tabs. There is no associativity between the AutoCAD and Inventor drawings; the result in AutoCAD is effectively a snapshot of the Inventor drawing.

2. **Inventor DWG** — the Save As command exports drawings in Inventor's dialect of DWG. Inventor objects are stored as blocks in layout tabs only. Model tab is empty.

   Inventor DWGs are fully associative back to the Inventor parts. If you modify the parts in AutoCAD, then the 2D views in the DWG file will update.

Because Inventor can save its drawings in DWG format, it is not necessary to use its native IDW format for drawings. After the Inventor drawing is in AutoCAD, you can perform minor editing, such as moving, deleting, and exploding.

Viewport blocks cannot, however, be modified, moved, copied, deleted, or even selected, but you can snap to geometry within them. Other Inventor blocks, such as borders and title blocks, can be edited. AutoCAD editing, such as adding text and other 2D geometry, is allowed.

AutoCAD does not have an "import Inventor" command.

While Inventor drawings cannot be exported to AutoCAD through the Clipboard, it is technically feasible to insert Inventor files as OLE objects in AutoCAD — technically feasible but not practical, because the files appear as icons only, and so are useless (as illustrated below).

Why the Complexity?

While opening AutoCAD drawings in Inventor is simple, importing them is a complex job. When it comes to the number of different ways to import *.dwg* files, I count... no, forget that. There are too many permutations of options to count them all.

The reason for the complexity is the lack of a one-to-one correspondence between Inventor and AutoCAD, unlike, say, between AutoCAD and AutoCAD LT. The differences lie in the approaches taken by the two CAD programs towards creating drawings and supporting objects types.

AutoCAD takes a relaxed approach to drawings, letting you freely intermix 2D and 3D. It doesn't even impose a default layer structure on you! It calls to mind the drawled reproach, "H-e-y dude, what's the big deal?"

Inventor takes a rigorous approach. Assemblies are made of, but kept separate from, parts, which are kept separate from drawings and presentations. It imposes its layer structure on your drawings. It reminds one of the stern reprimand, "A place for everything, and everything in its place." The good news is that compliance with standards is automatic, and pretty much 100%.

EXPORTING INVENTOR -> AUTOCAD SUMMARY

Export as DWG

Use the Save As and Save Copy As commands to export Inventor file to AutoCAD:

Environment	AutoCAD DWG	Inventor DWG	Notes
Assembly	No	No	Workaround: use Save File As, save as SAT, import into AutoCAD with AcisIn; each part is imported as a separate AutoCAD solid body.
Part	No	No	Workaround: use Save File As, save as SAT, import into AutoCAD with AcisIn.
2D Sketch	No	No	...
3D Sketch	No	No	...
Drawing	Save Copy As	Save As	Options available.
Presentation	No	No	Workaround: use Save File As, save as DWF or raster, attach to AutoCAD.

Copy, Paste to AutoCAD

Copy objects in Inventor to the Clipboard, and then attempt to paste into AutoCAD:

Environment	Command	Notes
Assembly	...	Not available in Inventor.
Part	...	Not available.
2D Sketch	Copy	Does not paste in AutoCAD.
3D Sketch	...	Not available.
Drawing	Copy (views only)	Does not paste.
Presentation	...	Not available.

IMPORTING AUTOCAD -> INVENTOR SUMMARY

Open | Import

Use the Open command's Import option to convert AutoCAD 2D and 3D objects into Inventor:

Import Option	Destination for AutoCAD Objects
3D Solids	3D solids placed in .ipt Part files, separate from 2D objects.
New Drawing	2D objects extracted and placed in Drawing environment.
Title Block	Title blocks extracted and placed in Drawing environment.
Border	Border objects extracted and placed in Drawing environment.
Symbols	Sketched symbol objects extracted and placed in Drawing environment.
New Part	2D sketch entities in Drawing environment;
	3D solids share same part file when **Solids to Single Part File** option is on.

Insert AutoCAD File

Use the Insert AutoCAD File command to import AutoCAD objects to Inventor:

Environment	Insert DWG	Notes
Assembly	Yes	MDT flavor of DWG files only, through Insert	Place Component command.
Part	No	IGES and SAT files only.
2D Sketch	Yes	2D objects only.
3D Sketch	Yes	2D objects only.
Drawing	No	…
Edit	Yes	2D objects only.
Presentation	No	…

Copy, Paste to Inventor

Copy objects in AutoCAD to the Clipboard, and then attempt to paste into Inventor. This method is primarily meant for copying selected 2D objects from AutoCAD to Inventor:

Environment	Command	Notes
Assembly	Paste Special	Command available, but no result.
Part	Paste Special	Command available, but no result.
2D Sketch	Paste	2D geometry pastes as sketches;
		3D geometry pastes as solids; blocks paste as blocks.
3D Sketch	Paste Special	Command available, but no result.
Drawing	Paste	Pastes 2D and 3D geometry as uneditable objects in a view; 3D geometry flatted to 2D.
Presentation	…	Not available.

Similarly for objects. While the two share the same solid modeling kernel and some 2D objects, differences exist. Inventor can only simulate AutoCAD's polyline-based rectangle, and can't cope with variable-width polylines at all. In turn, AutoCAD is unable to read Inventor's 3D models.

To deal with these yawning differences, Autodesk is forced to compromise the conversion process by providing two systems:

- Want a perfect looking conversion? Don't import; view only.

- Want an editable conversion? Import, but don't expect perfect results.

And that's just the way things are. The problem of converting drawings between different CAD systems is vexing, one that no organization has solved in thirty years of trying. Autodesk cannot be blamed for the situation; it troubles all CAD vendors, and is profitable for file conversion specialists. Autodesk has hinted that the AutoCAD and Inventor file formats may merge in the future, eliminating the viewing/importing dilemma.

At it best, importing is "pretty clean"; at its worst, there are pitfalls to be aware of. To help clarify matters, I've summarized the translation capabilities between Inventor and AutoCAD in the text boxes on the two previous pages.

Tutorial: Importing DWG Files to Inventor

Inventor can import AutoCAD drawings for editing. *Importing* involves translating AutoCAD objects into Inventor objects. This process is not as straightforward as the viewing of drawings described in the previous chapter, because the translation process requires you to make some decisions.

Here are the steps involved:

1. In Inventor, select **Open** from the **File** menu. Notice the Open dialog box.

2. In the **Files of Type** droplist, choose "AutoCAD Drawing (*.dwg)".

   Select a drawing file, such as *66 Profiles.dwg* found in AutoCAD 2008's *\help\buildyourworld* folder. This drawing contains both 2D and 3D objects, and model and layout tabs — very useful for seeing how Inventor imports them.

   But don't click Open just yet.

3. Click **Options**. Notice that the File Open Options dialog box has two options:

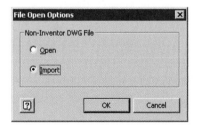

- **Open** — opens drawings for viewing, as described in Chapter 8.
- **Import** — translates AutoCAD objects into Inventor objects.

Choose **Import**, and then click **OK**

> **TIP** The default setting for Options is **Open**. This means you have to choose **Import** each time you want to import drawings; Inventor does not remember the previous setting.

4. Now you get to click **Open** and to start making decisions. Notice the first page of the import wizard.

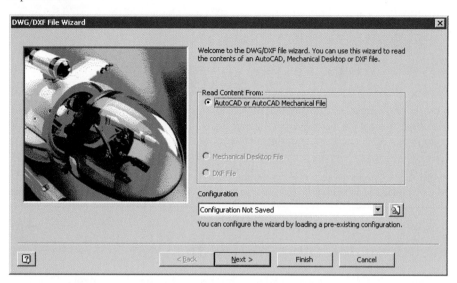

Read Content From usually figures out the file type automatically, but if not, then your choices are:

- **AutoCAD or AutoCAD Mechanical File** — .*dwg* files saved by programs like AutoCAD and Mechanical.

- **Mechanical Desktop File** — .*dwg* files saved by Mechanical Desktop, which require a different translator.

- **DXF File** — .*dxf* files saved by AutoCAD and other programs; DXF is short for "drawing interchange format," a file format invented by Autodesk long ago for accessing data stored in AutoCAD drawings.

Configuration — if you saved a configuration file from an earlier translation process, you can open it now. You have a chance to save the new configuration at the end of this wizard.

5. Click **Next**. Notice the Layers and Import Options page.

Selective Import — toggles the layers to import. The objects on each layer become separate sketches. For this tutorial, leave all layers on.

In other drawings, you probably don't want AutoCAD-generated layers like DefPoints and AShade, for example. Other unhelpful layers include those named Lights, Dimensions, and 0, if empty.

Selection | All — toggles whether the entire model is imported. To import just some objects, turn off this option, and then use the cursor to pick individual items in the preview window. (Try it: it works!) The preview window is actually a mini version of AutoCAD.

You can also use crossing and window selection modes:

- Window selection — drag from left to right

- Crossing selection — drag from right to left

- Shift+select — unselects individual objects; press **Esc** to unselect all objects.

When you cannot select objects, the cursor shows a padlock icon, as illustrated by the figure below of layout mode.

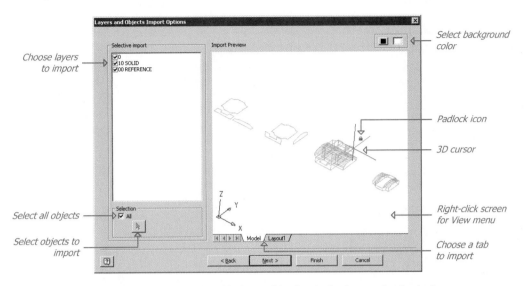

Choose layers to import

Select all objects

Select objects to import

Select background color

Padlock icon

3D cursor

Right-click screen for View menu

Choose a tab to import

Background Color — choose either black or white for the background color in the preview window. This does not override the background color when the AutoCAD drawing was saved with a visual style or workspace that overrides its background color.

Layout Tabs — choose the Model tab or a layout tab. Choose Model for this tutorial. Only the one tab you choose is imported into Inventor; the others are ignored.

TIP *Right-click the preview image for a shortcut menu that gives you access to viewing commands:*

You can use this to "straighten" out 3D views in model tab, or to choose a specific viewpoint, such as the top view of 3D models. When in Free Orbit mode, this shortcut menu becomes available:

IMPORT DESTINATION OPTIONS DIALOG BOX SUMMARY

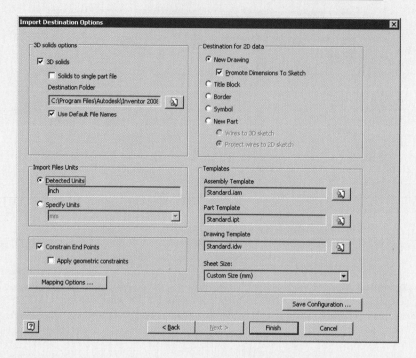

3D Solids Options

☑ **3D Solids** — imports sold models; turn off this option to omit importing 3D solids.

    ☐ **Solids To Single Part File** — imports solids into a composite feature; turn off this option to import solids as a single part file.

    Destination Folder — specifies the location for new Inventor files created from the imported solids. This option is unavailable when the Solids To Single Part File option is on.

        Browse — selects the folder.

    ☑ **Use Default File Names** — uses the name of the *.dwg* file to assign names to new files; when this option is unselected, you have to specify the file name the first time you save the file(s).

Import Files Units

⊙ **Detected Units** — reads the value of AutoCAD's InsUnits system variable (block insert units); when unitless, Inventor reads the value of the Measurement system variable, which specifies inches or millimeters.

○ **Specify Units** — selects units from the droplist.

☑ **Constrain End Points** — applies endpoint constraints automatically ; when off, you'll need to apply constraints manually in the sketch.

    ☐ **Apply Geometric Constraints** — constrains the sketch fully; when off, it doesn't.

continued

Mapping Options — maps layers and fonts in the *.dwg* file.

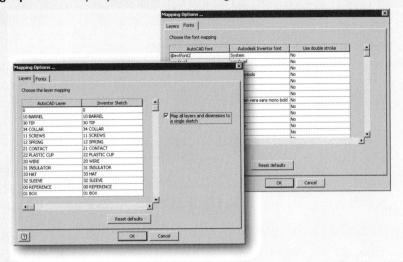

Destination For Data

⊙ **New Drawing** — creates a new *.idw* Inventor drawing file, and then places 2D data from specified layers.

☐ **Promote Dimensions to Sketch** — places the AutoCAD dimensions as associative sketch dimensions; when off, dimensions become drawing dimensions, which don't control features.

○ **Title Block** — places 2D data from specified layers into a title block.

○ **Border** — places 2D data from specified layers into a border.

○ **Symbol** — places 2D data from specified layers into sketched symbols in Drawing Resources, and then inserts instances of the sketched symbols on the first sheet.

○ **New Part** — translates 2D geometry as sketches. Note that associative dimensions are added to the sketch, but that unassociative dimensions, symbols, and other annotations are not imported.

○ **Wires to 3D Sketch** — places all wireframe data in 3D sketches.

⊙ **Project Wires to 2D Sketch** — projects wireframes to 2D sketches.

Templates

Assembly Template — specifies the name of the template for new assemblies.

Part Template — specifies the template for new parts.

Drawing Template — specifies template for new drawings.

Sheet Size — specifies sheet size from the droplist.

Save Configuration — saves settings to a configuration file, which can be reused for importing more *.dwg* files.

6. Click **Next**. Notice the Import Destination Options dialog box. I'll describe its options in detail later in this chapter; there's a summary in the boxed text on the previous pages.

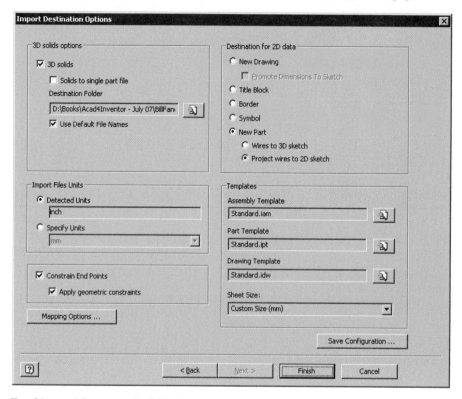

For this tutorial, turn on the following options:

- 3D Solids.
- Constrain Endpoints.
  - Apply Geometric Constraints.
- New Part.

The remaining options do not matter to this tutorial.

7. Click **Finish**. Notice that Inventor creates two files:

- 2D objects are placed as sketches in a part file.
- 3D objects are placed as parts in an *.iam* assembly file.

(In some cases, you might see a blank *.idw* drawing file; close it in order to see the assembly.)

> **TIP** You can also insert 2D .dwg files into existing sketches with the **Insert AutoCAD File** button on the 2D Sketch panel.

8. Use the **Window** menu to switch between "Part1" and "66 Profiles.iam." The imported 2D objects are sketches in a part file, and look like this:

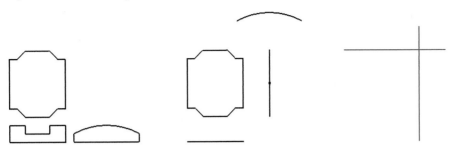

The two lone black lines and arc are 2D objects that drawn vertically in AutoCAD, and then projected to 2D by Inventor's translation. The two red lines are construction lines hidden by the 3D models. The original AutoCAD drawing is shown below.

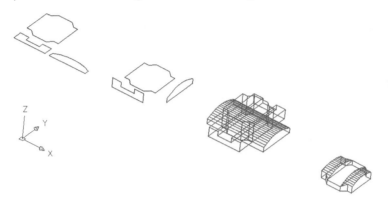

The imported 3D objects are parts in an assembly file, and look like this:

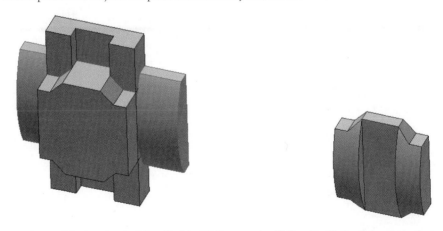

The imported AutoCAD drawing is fully editable. Well, not quite "fully editable," rejoins the technical editor: each AutoCAD solid becomes a separate *.ipt* part file. You can move parts, edit materials (finishes), and constrain parts. But, you cannot edit them in the context of the assembly; you need to open each one in the Part environment for editing.

In summary, the following occurs by default when importing .*dwg* files (Note that there are some differences, depending on whether the drawing is imported into a 3D part file or 2D drawing file.)

- 2D geometry is placed in one or more sketches attached to a draft view.

- Dimensions are added as associated geometry to the same sketch.

- Blocks are converted to sketched symbols in the Drawing Resources folder in the browser. Dynamic blocks lose their unique features.

- Unassociated dimensions, symbols, and other annotations are placed on a drawing sheet.

- Solid models become separate .*ipt* part files, one per part. An .*iam* assembly file is created to reference the new part files. Note that Inventor uses default file names for the parts. Do not rename them with Windows Explorer, because then they lose their connection to the assembly. To rename parts within an assembly, use Inventor's Design Assistant.

- To edit parts, you need to install the experimental Feature Recognition module from Autodesk Labs (labs.autodesk.com). Open a part file, and then apply the **Feature Recognition** command to break parts down to their constituent features and underlying sketches. These can now be edited. When you are finished editing, save the changes, and close the part file; the assembly will update.

- Text in tables can be edited.

TIP *Blocks can be edited directly using the grips illustrated in the figure below.*
- *Green grips — move block.*
- *Blue grips — rotate block.*
- *Yellow grips — resize block.*

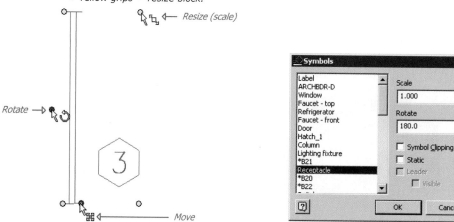

To change a block, double-click it, and then select the name of another block from the Symbols dialog box.

TIP *When the following dialog box appears during AutoCAD insertion, you need to install a "hotfix" from Autodesk.*

You can download the 154.6MB hotfix TS1071050 from usa.autodesk.com/ getdoc/id=DL9741651, and then follow these steps:

1. *Using the Control Panel's **Add or Remove Programs** applet, uninstall AOemView 2008 (AutoCAD OEM).*
2. *Unzip the TS1071050.zip file using the **All files/folders in archive** and **Use Folder Names** options.*
3. *Double-click **Setup.exe** to start installation. The routine reinstalls AOemView 2008 and its support files.*
4. *Following installation, click **Finish** to complete the install and exit the dialog.*

Design Doctor Recovers Import Problems

During importing, you may need repeatedly to tell Inventor the location of *.shx* files, such as *simples.shx* and *ltypeshx.shx*. Click **OK** until you get tired of doing it, then click **Cancel** for the importing process to continue. In this case, some fonts may be replaced by the font designated as the alternative by AutoCAD's FontAlt system variable, and complex linetypes may be missing their shapes.

Text that is styled to be backwards or upside-down is not displayed correctly. The overscore metacharacter (%%o) is reproduced literally.

Rays and Xlines are not displayed. DrawOrder is ignored. Traces and 2D solids are sometimes unfilled. Variable-width polylines have constant width.

When the drawing contains dimensions that aren't attached to objects, Inventor will warn you in a dialog box, as illustrated below.

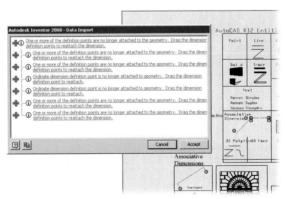

The red underlined warning messages are hyperlinked: click on a message and Inventor highlights the dimension and zooms in so that you get a closer look.

Click the bold red cross to fix the problem: Inventor asks if you wish to run the Recover command. Answer in the affirmative, and the Design Doctor makes his house call.

1. Select one problem to fix, and then click **Next**. (It fixes one problem at a time.)

2. Inventor zooms in to the problem object, and gives you chance to fix it manually. In this case, it's attaching the dimension to an entity. (When done, or if you'd rather that Design Doctor make the cure, click **Next**.)

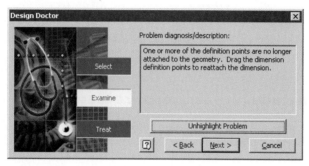

3. Design Doctor provides three possible cures, as illustrated below. In this case, they are (1) reattach the limb, (2) amputate, or (3) build a prosthesis. Choose one, and then click **Finish**.

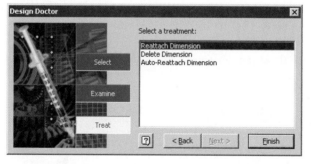

In other cases, Inventor may report an error message when blocks contain mtext. In this case, Inventor refuses to import the *.dwg* file. The suggested workaround is to use AutoCAD to explode blocks containing mtext, and then re-attempt the import.

Inventor cannot import *.dwg* files that are open in AutoCAD.

Tutorial: Exporting Inventor Files to AutoCAD

Going in the other direction, Inventor can export its drawings to AutoCAD:

- All of Inventor's layers names and properties, linetypes, text styles, and dimension styles are added to the *.dwg* file.

- AutoCAD objects opened in Inventor for viewing are returned intact.

There are, however, these limitations:

- Inventor drawings can be exported to AutoCAD only from its Drawing environment.

- Other than DWG, AutoCAD cannot read any of Inventor's native formats.

- All objects are exported as 2D; 3D views are flattened to 2D.

As I noted earlier, Inventor can export drawings in two variants of the DWG format. They share the similarities listed above, but differ from each other in the following manner:

AutoCAD DWG — use the Save Copy As command to export drawings in AutoCAD's flavor of DWG. When the *.dwg* file is opened in AutoCAD, you can expect the following results:

- Inventor objects are converted to individual AutoCAD objects, including AutoCAD objects that were earlier imported (converted) to Inventor format.

- 3D views lack rendering.

- All objects, include title blocks and borders, look the same in Model and layout tabs.

Inventor DWG — use the Save As command to export drawings in Inventor's flavor of DWG. In AutoCAD, expect the following results:

- Inventor objects are stored as editable blocks, including AutoCAD objects that were earlier imported (converted) to Inventor format. The exception are part views, which cannot be edited.

- 3D views lack rendering.

- Objects are placed in layout tabs only; model tab is empty.

Export in AutoCAD DWG Format

In this tutorial, you export an Inventor drawing in AutoCAD DWG format.

1. In Inventor, open the *Tire Rim.idw* drawing file from Inventor's *\samples\models\parts\tirerim* folder.

    This sample drawing is good for this tutorial, because it contains dimensions, 2D views, and a rendered 3D view.

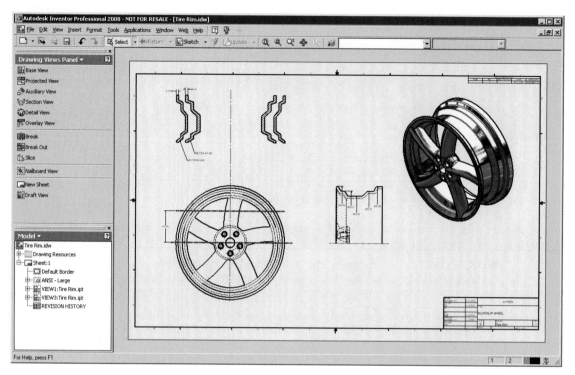

2. From the **File** menu, choose **Save Copy As**. This command differs from Save As in that it makes copies, leaving original drawings unchanged.

    Notice the Save File As dialog box.

3. In the File Save As dialog box, make these changes:

    *   Change **Save In** to a convenient place to locate the file, such as "C:\."

    *   From the **Save as Type** droplist, choose "AutoCAD Drawing (*.dwg)."

    *   In the **File Name**, change the name to "Tire Rim-acaddwg.dwg," which helps you distinguish it from other versions of the same drawing. (Acaddwg is code for "saved in AutoCAD's DWG format.")

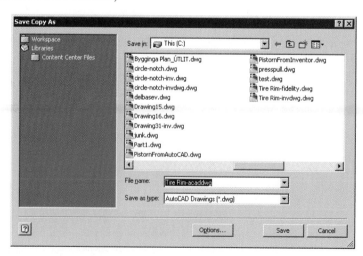

4. Click **Options**. Notice the DWG/DXF Export Options dialog box.

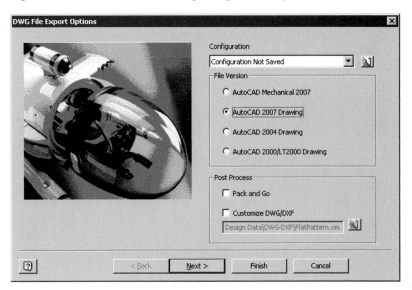

5. In the DWG/DXF Export Options dialog box, make the following changes for this tutorial:

    • Under **File Version**, choose "AutoCAD 2007 Drawing."

    • Ensure that **Pack and Go** and **Customize DWG/DXF** boxes are unchecked (turned off).

    (Later in this chapter, I'll explain in detail all of the options in this dialog box.)

6. Click **Next**. Notice the Export Destination dialog box. For this tutorial, leave options at their default settings, as illustrated below.

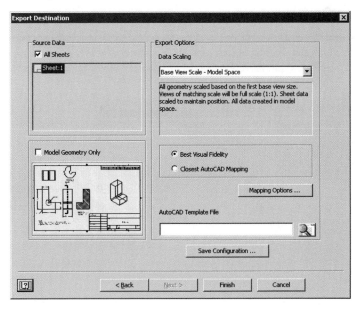

7. Click **Finish**. Notice that the Save Copy As dialog box reappears.

8. Click **Save**. If a message appears warning about project paths, ignore it by clicking **OK**.

9. Open *Tire Rim-acaddwg.dwg* in AutoCAD. It should look like the figure below.

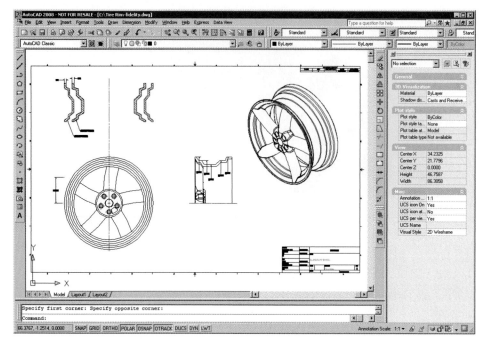

10. Check out the results of the Inventor->AutoCAD translation by opening the Properties palette, and then selecting objects to see what they are made of.

   • Dimensions are dimensions on layer "Dimension(ANSI)."

   • Hatch patterns are hatch patterns on layer "Hatch(ANSI)."

   • Entities are made of lines, arcs, circles, and polylines on layer "Visible(ANSI)."

   • Leaders and tables, however, are blocks, rather than mleader and table objects.

   • The title block and drawing border each are blocks on layer 0. When you open the title block in the block editor, you'll find that the text consists of attributes (see figure below).

Exporting in Invent's DWG Format

To see the difference between AutoCAD's and Inventor's DWG formats, repeat the exercise.

1. This time, however, use the **File | Save As** command to save the drawing as "Tire Rim-invdwg.dwg" in Inventor Drawing (*.dwg) format.

   Notice that there are no translation options. Ignore the project path complaint.

2. Open the drawing in AutoCAD. Notice that it looks more like the original in Inventor, because the 3D view is now shaded.

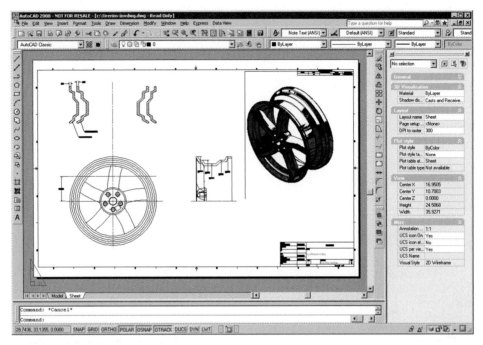

3. Check out the other differences:

   • Everything is grouped into blocks.

   • You cannot select view blocks. (You can, however, access them through the BEdit command, as illustrated below.)

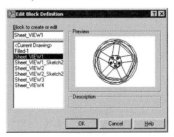

   • Blocks you can select can be exploded, and then edited.

   • Model tab is empty.

Comprehensive Import Destination Reference

If importing an AutoCAD drawing fails to turn out as you expect, the problems may be due to incorrect settings in the Comprehensive Import Destination dialog box. The following sections provide the details, along with a brief explanation of the significance of many options.

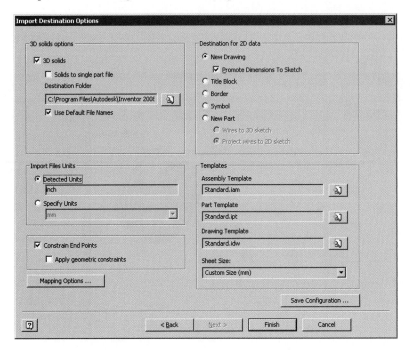

3D Solids Options

These options determine if and how 3D solids are imported.

3D Solids — imports AutoCAD solid models. Each model becomes a separate Inventor *.ipt* part file; Inventor then creates an *.iam* assembly file that references the parts.

☑ imports 3D solids.

☐ excludes 3D solids from being imported.

Why? If the AutoCAD drawing is primarily a 3D solid model, then turn on this option. If you only want the 2D data from the drawing, such as the title block, then turn off this option to avoid importing 3D solids.

> **Solids to Single Part File** — determines how solids are imported; this option works correctly only when the Destination for 2D Data is set to "New Part."
>
> ☑ imports AutoCAD drawing as a composite feature — combines 2D and 3D objects into a single part file.
>
> ☐ imports all solids to a separate file — places 2D and 3D objects in separate files.
>
> *Why?* This option lets you decide whether you want 2D objects separate from 3D ones, or combined in a single part file. If separate (option turned off), then Inventor creates a *.ipt* part file for 2D (imported as sketches), and an *.iam* assembly file for the 3D solids.
>
> When this option is turned on, a single part file is created that contains both 2D (as sketches) and 3D objects.

Destination Folder — specifies the folder for storing Inventor files created from imported AutoCAD solids. The default folder depends on the project file; for *tutorial.ipj* it is *c:\program files\autodesk\inventor 2008\tutorial files/*.

Why? By specifying a different folder, you can segregate the imported files from other files.

Use Default File Names — assigns names to new Inventor files based the *.dwg* file name; the name can be changed upon saving. Change the file name:

☑ assigns default file names.

☐ requires the user to specify file names upon saving.

Why? If you go for the defaults, and then rename the files later using Windows Explorer, you break the links between the files, which is undesirable.

Import Files Units

These options determine how Inventor assigns units to imported drawings. This is necessary, because AutoCAD drawings are unitless; the units you set with the Units command are for display purposes only.

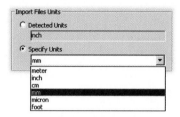

⦿ **Detected Units** — reads the value of the InsUnits system variable stored in the AutoCAD drawing, and attempts to match it to one of Inventor's units. (InsUnits specifies the units for inserted blocks).

If the *.dwg* file is unitless or set to units not supported by Inventor, then it reads the value of the Measurement system variable to determine whether the drawing should be Imperial units or metric. (Measurement specifies inches or mm.)

○ **Specify Units** — selects a conversion unit from the droplist. (See comparison table in the boxed text.)

Why? If units are set incorrectly, Inventor may get sizes and scale factors incorrect, which can affect analysis and drawing.

Constrain End Points — determines whether endpoint constraints are applied to AutoCAD objects imported as sketches:

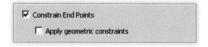

⦿ endpoint constraints applied automatically.

○ endpoint constraints not applied.

Why? Inventor draws some objects differently from AutoCAD. For example, the rectangle is made of four lines with end constraints. The constraints keep the rectangle together, making it appear as a single object. (AutoCAD uses a closed polyline for a rectangle, and so its rectangle truly is a single object.) By applying end constraints, the converted drawing acts like the original did in AutoCAD.

You might want to turn off this option if Inventor constrains endpoints in ways you don't want. On the other hand, if you are using the objects as sketch geometry to define solid features, then the endpoints *must* be constrained to one other.

AUTOCAD-INVENTOR UNITS SUMMARY

InsUnits	Meaning in AutoCAD	Available in Inventor
0	No units	...
1	Inches	inch
2	Feet	foot
3	Miles	...
4	Millimeters	mm
5	Centimeters	cm
6	Meters	meter
7	Kilometers	...
8	Microinches	...
9	Mils	...
10	Yards	...
11	Angstroms	...
12	Nanometers	...
13	Microns	micron (millionth of a meter)
14	Decimeters	...
15	Dekameters	...
16	Hectometers	...
17	Gigameters	...
18	Astronomical Units	...
19	Light Years	...
20	Parsecs	...

Apply Geometric Constraints — determines whether to apply all possible constraints to the imported sketch:

⊙ geometric constraints applied.

○ geometric constraints not applied.

Why? The Constrain End Points option applies only endpoint constraints to the imported drawing. If you want Inventor to apply all possible constraints, then turn on this options. Other constraint types include concentric, colinear, perpendicular, and tangent.

You would leave this option turned off, because Inventor can add so many constraints that they become unmanageable. On the other hand, if you are turning sketch geometry into solids, then it should be constrained. Automatically-applied constraints can be deleted if they don't suit your purpose.

Mapping Options — opens the Mapping Options dialog box for matching layers and fonts between AutoCAD and Inventor. Described later in this section.

[Mapping Options ...]

Why? Match layers when it is important that AutoCAD objects be assigned to Inventor's preset layer names.

You want to match fonts when the Inventor installation lacks the fonts used in the AutoCAD drawing. There are a couple of reasons why Inventor might not have access to the source fonts: (1) when the AutoCAD drawings come from another computer, which has a different collection of fonts, and (2) lookalikes may have to be substituted when the source fonts are covered by copyright. All fonts included with the Windows operating system may be freely copied.

Destination For 2D Data

The Destination set of options specify how you want Inventor to translate 2D objects from AutoCAD drawings.

This is probably the most important part of this import wizard, so I'll show examples of how each option affects the following AutoCAD 2008 sample drawing, *Mechanical - Multileaders.dwg* (found in the \sample\mechanical samples folder). The drawing is of wellhead piping.

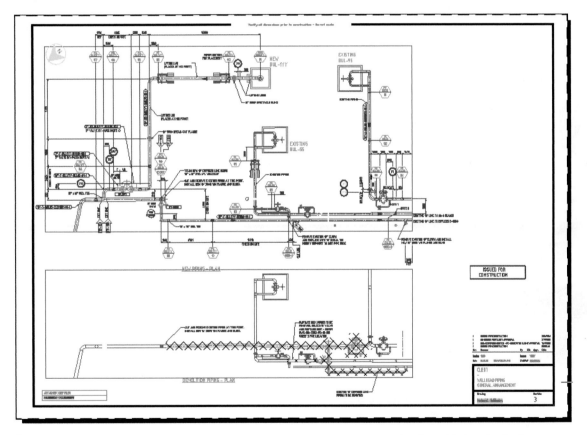

⊙ **New Drawing** — converts 2D geometry and dimensions into one or more sketches added to the draft view of an *.idw* drawing file. Annotations and unassociated dimensions are included on the drawing sheet.

Xref'ed blocks are bound and, together with blocks, converted to sketched symbols, and then stored in the Model browser's Drawing Resources folder. See figure below.

Why? Use this option when you want to convert AutoCAD 2D drawings and blocks into Inventor 2D drawings and symbols.

Promote Dimensions to Sketch — determines how to handle dimensions:

☑ converts dimensions to associative *sketch* dimensions, which act like AutoCAD dimensions: they change when the attached sketches are moved or edited. Dimensions control features when imported into the Part environment, but not in the Drawing environment.

☐ converts dimensions to non-associative *drawing* dimensions, which are not attached to sketch objects. When objects are moved, these dimensions do not update.

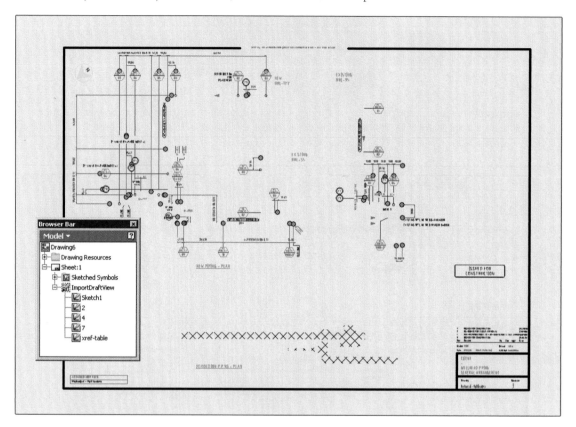

TIPS *Draft view is like an AutoCAD drawing. It can be empty, can contain multiple sketches, and can be scaled. (Inventor drawings are otherwise always 1:1.) Inventor always places imported AutoCAD 2D drawings and their dimensions, text, and other annotations on a drawing sheet in draft view.*

Whether the contents of model space or a layout tab appear depends on which you selected in page 2 of the import wizard.

If the Recover dialog box appears, agree to its recommendations by clicking **Accept**; *otherwise the imported drawing may be blank.*

○ **Title Block** — converts the AutoCAD drawing into a *.idw* drawing file, placing all 2D data in the Model browser's Title Blocks section, as illustrated below. While blocks are imported, they are not segregated as sketch symbols.

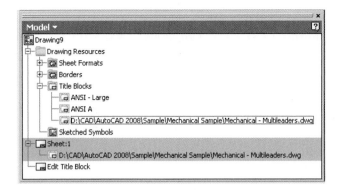

Why? Use this option to import title blocks from AutoCAD. There's less cleanup when the *.dwg* files contain just the title block, and no other objects.

○ **Border** — converts the AutoCAD drawing into an *.idw* drawing file, placing 2D data into the Model browser's Borders section.

Why? Use this option to import drawing borders from AutoCAD.

○ **Symbol** — converts an AutoCAD drawing into an *.idw* drawing file, placing all 2D data into a single sketched symbol in the Model browser's Sketched Symbols section. The drawing is inserted as a single block-symbol in Sheet:1 of the drawing.

Why? Use this option to import drawings as block-symbols. Unfortunately, the entire drawing becomes one symbol, which makes this less than useful for block library drawings.

○ **New Part** — converts 2D objects in the AutoCAD drawing into a sketch in an *.ipt* part file; from model space or paper space. Associative dimensions and some text are included, but unassociated dimensions and other annotations are not imported.

Why? Use this option to import drawings as sketches, features, and objects. These can then be edited to 3D parts through extrusion, revolving, and so on.

There are two additional options for "wires," automotive jargon for wireframe meshes. These options are meant for Alias drawings exported in DWG format.:

○ **Wires to 3D Sketch** — places wireframe data in a 3D sketch.

⊙ **Project Wires to 2D Sketch** — flattens wireframes to 2D sketches.

Templates

Determines which template files are used for imported AutoCAD drawings. Click the **Browse** button to choose different files.

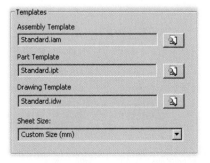

Assembly Template — specifies the *.iam* template file for new assemblies.

Part Template — specifies the *.ipt* template file for new parts.

Drawing Template — specifies the *.idw* template file for new drawings.

Sheet Size — selects the default sheet size for drawings.

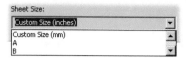

The range is A - F (American sizes), A0 - A4 (metric size), and Custom Size in inches or millimeters. The Custom Size option sizes the sheet to fit the drawing.

Why? If you don't know (or don't care) about the sheet size, then select Custom Size. If the drawing must fit a specific sheet size, then select its size-name from the droplist. "A" represents letter-size paper, "B" is double the size of A, and so on.

Save Configuration — saves these settings an *.ini* configuration.

Why? If you have more AutoCAD drawings to import, you can reuse all these settings without having to select them over and over again.

Comprehensive (Import) Mapping Options Reference

Normally, Inventor converts AutoCAD layers one for one, adopting the same names and properties — with the exception of those properties not supported by Inventor, such as frozen viewports. The Mapping Options dialog box lets you to override Inventor's choices in mapping AutoCAD layer and font names.

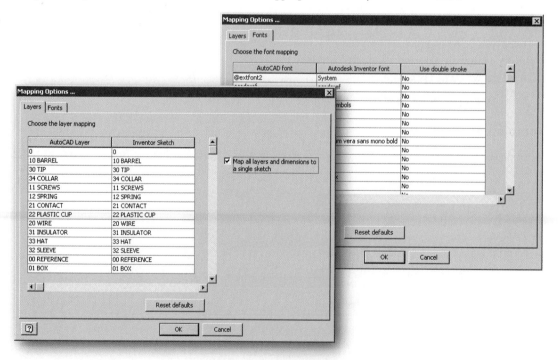

In addition, Inventor segregates the content of layers into individual sketches, as illustrated below. For example, everything on layer 00Border becomes a sketch named "00Border." After importing the drawing, you'll see layer 00Border and sketch 00Border in Inventor.

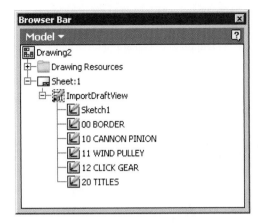

The Mapping Options dialog box lets you combine the content from AutoCAD layers to create fewer sketches.

TIP *Layers turned off or frozen in .dwg files are not imported.*

Inventor also attempts to match AutoCAD fonts to ones in Inventor, but it's not quite as easy as with layers. For example, some font files on a computer running AutoCAD may not exist on another computer running Inventor. And Inventor doesn't support Autodesk's SHX font format at all, substituting identically-named TrueType (*.ttf*) fonts. (Autodesk provides matching TrueType fonts with both AutoCAD's and Inventor's installation.)

On the Layer tab, we have the following options:

AutoCAD Layer — lists the names of layers in the AutoCAD drawing.

Inventor Sketch — lists the layer and sketch names Inventor will be assigning. Click a name to reveal a droplist of all layer names, and then choose another one; this somewhat awkward method lets you merge layer content.

Map All Layers and Dimensions To A Single Sketch — is the quick way to get everything onto a single layer and into a single sketch named "Sketch1." (The "unused" layer names are still imported and are empty.) Autodesk warns that mapping large amounts of data to a single sketch may increase conversion time.

Reset Defaults — restores mapping to its original setting, one for one.

And on the Fonts tab, we have these options:

AutoCAD Font — lists the names of fonts provided by AOEMView, Autodesk's conversion engine used by Inventor. (It is installed at the same time as Inventor.) Unfortunately, fonts used by the imported drawing are not segregated from the 72 others, so you don't easily know which need attention.

At the bottom of the list is "Unknown/Missing Font," to which you can map a single font for all missing fonts. A good one to use is Arial or Tahoma.

Inventor Font — lists the nearest equivalent font available to Inventor. All Autodesk and Windows fonts installed on your computer are listed, so this list gets really long. To change the mapping, click a name to reveal a droplist of all fonts, and then choose another one.

AutoCAD font	Autodesk Inventor font	Use double stroke
isoct2	isoct2	No
isoct3	isoct3	No
italic	Gothic ITC Heavy Italic BT ▼	No
italic8	Franklin Gothic ITC Heavy It ▲	No
italicc	FranklinGotTDemCon	No
italict	Futura Black BT	No
	Futura Bold BT	
ltypeshp	Futura Bold Condensed BT	No
monotxt	Futura Book BT	No
monotxt8	Futura Extra Black BT	No
romanc	Futura Extra Black Condens‹	No
romand	Futura Extra Black Condens‹	No
	Futura Extra Black Italic BT	
romans	Futura Heavy BT	No
romant	Futura Light BT ▼	No
scripts	_scripts_	_No_

Use Double Stroke — doubles the strokes (lines) to fonts, making some fonts thicker and thus easier to read. Double-stroking is best for thin SHX fonts.

Use Single Stroke Font — resets all fonts to use single stroke.

Reset Defaults — restores mapping to its original setting, one for one.

Comprehensive DWG/DXF Export Reference

In Inventor's Drawing environment, you can export drawings in AutoCAD's DWG or DXF formats. There are a number of options available for controlling the export process, as described here.

TIPS *These options are available only when exporting drawings and sheet metal patterns in AutoCAD's DWG or DXF format:*

- *To export a drawing: from the **File** menu choose **Save Copy As.***
- *To export a sheet metal pattern: right-click the pattern, and then choose **Save Copy As** from the shortcut menu.*

*To access these options, click the **Options** in the Save File As dialog box.*

These options are unnecessary when saving drawings in Inventor's DWG format.

Export as DWG files when the target is AutoCAD or AutoCAD LT; export as DXF files when the target is a non-Autodesk CAD or CAM program.

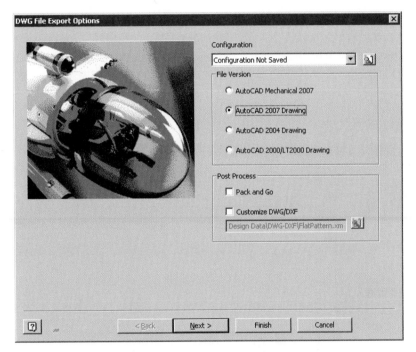

Configuration — selects a previously saved configuration file. Click **Browse** to find an *.ini* file not shown by the droplist. If you have no need to change settings after loading the configuration file, click **Finish**.

Why? Configuration files adjust all export options to settings you or anyone else chose on a previous occasion. A CAD manager can create configuration files for use by drafters and clients, standardizing the export process. You create configuration files as the last step in this export process by clicking the **Save Configuration** button.

File Version

The File Version option selects the release (version) number for the drawing file. Choose a version number that best matches the release of AutoCAD that will be reading this exported file.

The version numbers listed as options represent the major changes in the DWG format:

- ○ **2000** is also read by AutoCAD 2000i, 2001, and 2002, as well as LT.
- ◉ **2004** by AutoCAD 2004, 2005, and 2006 (default setting).
- ○ **2007** by AutoCAD 2007, 2008, as well as the upcoming 2009.

Why? This option is needed in case the destination release of AutoCAD is not 2008. If you are not sure which version to choose, then select **2000**, because it works with the most copies of AutoCAD. There is a drawback, however, to using a version number earlier than necessary: there is a chance that some objects might be lost or altered by conversion – although I have not verified if this occurs.

(If the export format is DXF, then you have the following choices: AutoCAD Mechanical 2007, AutoCAD 2007 DXF, AutoCAD 2004 DXF, AutoCAD 2000/LT2000 DXF, and AutoCAD R12/LT 2 DXF.)

Pack and Go

This option places the *.dwg* drawing and *.shx* files in a *.zip* file; font and non-font *.shx* files are placed in a folder named *fonts*. This is a simplified version of AutoCAD's eTransmit command.

- ☑ after clicking Save, the *.zip* file is created.
- ☐ *.zip* file is not created,

The *.zip* file has the same name as the drawing. For example, the drawing from *TireRim.idw* is stored in *TireRim.zip*. You cannot open *.zip* files with AutoCAD; instead, you must first extract the contents of the *.zip* file.

Windows Explorer in XP and Vista read *.zip* files and extract their content; Windows 2000 needs a program like WinZip to retrieve the content.

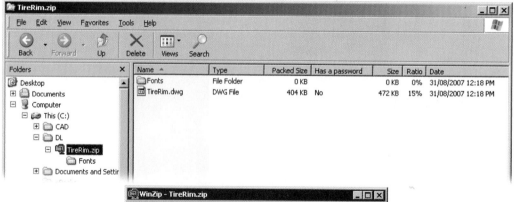

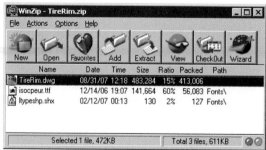

Top: Windows XP's Explorer displaying contents of .zip file
Above: WinZip displaying contents of .zip file.

Why? Except for the purposes listed below, you would tend to leave this option turned off. This option is useful for these purposes:

- Provides a list of the fonts used by the drawing, and includes the fonts files.

- Collects drawings and support files for archival purposes.

- Compresses drawings for transmittal by email. In the figure above, the drawing's size has been compressed by 15%.

Customize DWG/DXF

Turning on the Customize DWG/DXF option controls the export process according to settings defined in an *.xml* file:

☑ changes the exported drawing according to the settings in the selected *.xml* file.

☐ exports drawings as they are.

Click the **Browse** button to choose an *.xml* file. Some are available in Inventor's *design data**dwg-dxf* folder. It seems that the only way to create new customization files is through hand-coding: open a file like *FaceLoops.xml* in Notepad, make changes and additions, and then save by another name.

Autodesk's documentation says this option merely "defines layer names," but it does much more, as a reading of the sample *.xml* files shows:

- **setName** — changes the names of layers.

- **deleteLayer** — deletes layer names and the content of the layers. The exception is layer 0: it cannot be deleted, only its content.

- **replace** — changes entities, such as the chord tolerance of splines.

These settings were originally designed to process sheetmetal designs for other programs. The settings tweaked Inventor's output to make the result more compatible with other software.

Comprehensive Export Destination Reference

After clicking **Next** in the DWG File Export Options dialog box, the Export Destination dialog box appears:

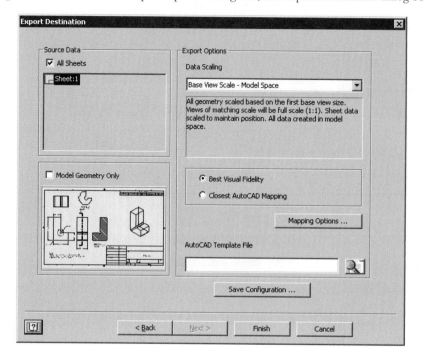

Source Data

The Source Data option selects the sheets to export. Inventor's sheets are equivalent to AutoCAD's layouts.) The meaning of the All Sheets option is:

☑ all sheets are exported.

☐ only the sheets selected in the list below are exported.

Why? This option exports some or all sheets from the Inventor drawing. If you are unsure, choose the All Sheets option.

> **TIP** *Icons next to sheet names report their status, such as out of date or missing links. You can still export the drawing, even with these warnings in effect.*

Model Geometry Only

The Model Geometry Only option exports only model geometry, specifically the drawing's model, base, and projected views.

☑ only model geometry is exported.

☐ all geometry in the drawing is exported.

Why? This option is useful for cleaning up the exported drawing by excluding non-model geometry, such as dimensions, drawing borders, title blocks, and annotations. If you are not sure, turn off this option to export everything; you can always erase unneeded objects in AutoCAD.

The preview image associated with this option illustrates the difference:

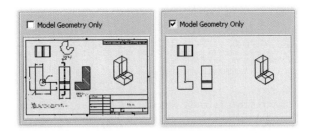

Left: *Entire drawing is exported.*
Right: *Only model geometry and associated views are exported.*

Data Scaling

The Data Scaling options lets you choose the scale factor (base, sheet, or full scale) and destination (model or layout) for exported drawings.

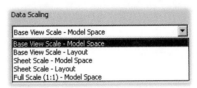

The options are:

Base View Scale - Model Space — scale factor of the first base view (viewport) is taken as 1:1; other geometry and sheet size are scaled to match. All geometry is placed in Model tab.

Base View Scale - Layout — scaling as above. All geometry is placed in paper space.

Sheet Scale - Model Space — sheet keeps its size and geometry is not resized.

Sheet Scale - Layout — as above.

Full Scale (1:1) - Model Space — all geometry is resized to full scale. As a result, some viewports may overlap in AutoCAD.

Best Visual Fidelity & Closest AutoCAD Mapping

These options fine-tune the exported geometry either to look nicer or be more accurate.

Best Visual Fidelity — uses AutoCAD geometry that *looks* most like the Inventor drawing.

Closest AutoCAD Mapping — uses AutoCAD geometry that best approximates the Inventor drawing.

Why? Recall that Inventor and AutoCAD don't have an identical set of equivalent objects, and so compromises must be made.

Mapping Options

This option opens the Mapping Options dialog box, as described later. It maps linetypes, sets line type scaling, and determines how dimensions and symbols are handled.

AutoCAD Template File

Select a *.dwg* or *.dwt* file for use as a template in creating the exported AutoCAD drawing. Click the **Browse** button to choose the file.

Why? Inventor reads the layer properties from the template file for the exported file.

Save Configuration — saves these settings as an *.ini* configuration file.

Why? If you have more AutoCAD drawings to export, you can reuse these settings without having to select them over again. A typical *.ini* file looks like this:

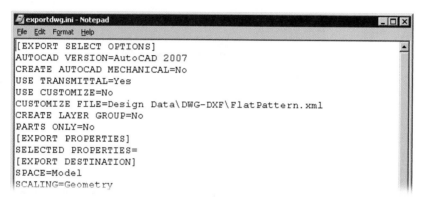

Comprehensive Mapping Options Reference

The Mapping Options maps linetypes, sets line type scaling, and determines how dimensions and symbols are handled. The default values are the closest possible matches between Inventor and AutoCAD, and so you may never need to bother with this. To access this dialog box, click **Mapping Options** in the Export Destination dialog box.

Click the **Line Types** tab to see the following options:

> **Inventor Line Type** — lists the names of all line types available in Inventor. Line types actually used by the drawing are not highlighted, unfortunately.
>
> **AutoCAD Line Type** — lists the line types included with... well, not AutoCAD. This list of linetypes is defined by the currently loaded *.lin* file, such as *invANSI.lin* in the figure below — which is Inventor's linetype file.
>
> In any case, to change a match, click a line type name, and then choose another from the list.
>
> **AutoCAD Line Scale** — changes the scale factor for each line type. To make changes, click "Same as Inventor" (the default), and then enter a decimal number, such as 0.5 or 2.

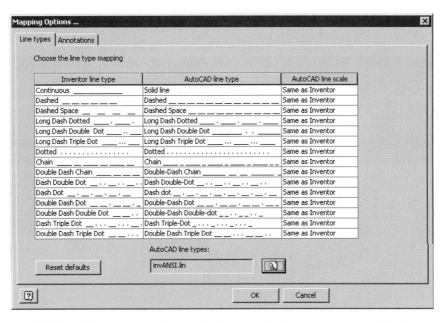

AutoCAD Line Types — specifies the *.lin* file to use for line type mapping. Click the **Browse** button to choose a different one. Unfortunately, when you choose another *.lin* file, this dialog box sets all matches to "Solid Line."

Reset Defaults — restores the default *.lin* file, which matches the drawing standard used by the drawing, such as ANSI, ISO, and DIN.

Click the **Annotations** tab to see the following options:

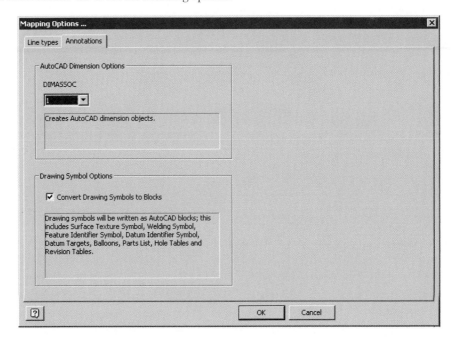

DIMASSOC — toggles the type of dimensions converted to AutoCAD:

1 creates non-associative dimensions.

0 creates exploded dimensions, which are made of lines, text, and arrowhead blocks.

Convert Drawing Symbols To Blocks — toggles how symbols in the Inventor drawing are handled:

☑ symbols are converted to blocks.

☐ symbols are exported as geometry.

Inventor's symbols are found in the Drawing environment's Drawing Annotation panel, and include:

- Surface texture symbols.
- Welding symbols.
- Feature identifiers.
- Datum identifiers.
- Datum targets.
- Balloons.
- Part lists.
- Hole tables.
- Revision tables.

What Works, and What Doesn't

Not every AutoCAD object, property, and table imports correctly into Inventor. The following tables provide a summary. Note that the results of the import differ depending on the environment into which the AutoCAD drawing is imported — parts or sketches.

Object	Open *	Import
3D solids	Yes	Yes
3D surfaces	Yes	Yes
3D faces	Yes	Yes
Arc	Yes	Yes
Attribute	Sometimes	Text only
AttDef	Sometimes	Yes
Blocks	Yes	Yes
Blocks, Dynamic	As static blocks	As static blocks
Circle	Yes	Yes
Dimension	Yes	Yes
Ellipse	Yes	Yes
Field text	As text	As text
Gradient	Solid filled	Solid filled
Hatch	Yes	Yes, but cannot edited
Helix	Yes	Yes
Hyperlinks	Text only	Text only
Image	Yes, if found	Yes, if found
Line	Yes	Yes
MLeader	Yes	Not sure
MLine	Yes	**No**
MText	Yes	Yes
OLE Object	Yes	Yes
Point	Yes	Yes
Polyline	Yes **	Yes, but variable width polylines display uniform width; donuts have incorrect width and become two arcs; segments and polyarcs become individual lines and arcs.
Proxy	Yes	Yes
Ray	**No**	**No**
Region	Yes	Yes
Section	**Cutaway shown,** no section object.	Yes, but appears as 3D solid box.
Shape	**No**	**No**
Solid (2D)	Yes	Yes, but some solids may be unfilled; borders become individual lines.
Spline	Yes	Yes
Table	Yes	Yes
Text	Yes	Yes, but some display errors.
Tolerance	Yes	Yes, but scale factor may be wrong.
Trace	Yes	Yes, but some traces may be unfilled; borders become individual lines.
XLine	**No**	**No**

Notes:
*) Basic entities, such as lines and arcs, drawn in 3D are shown in 2D plan view.
**) Variable-width polylines are not displayed correctly.

View Mode	Open	Import
Model space	Yes	Yes
MS Viewports	**No**	**No**
Layouts	Yes	Yes
PS Viewports	Yes	Yes
Viewpoint	Yes	Yes
DrawOrder	**No**	**No**

Table Entry	Open	Import
Annotative styles	**No**	**No**
Block names	Yes	Yes
Dimension style	Yes	Yes
Layer names	Yes	Yes
Linetype names	Yes	Yes
MLeader styles	Yes	Yes
MLine styles	**No**	**No**
Plot styles	**No**	**No**
Shapes	**No**	Yes, if *.shx* file found.
Solid fill	Yes	Varies
Table styles	**No**	**No**
Text styles	Yes	Yes
Visual style names	**No**	**No**

Property	Open	Import
Color	Yes	Yes
Simple Linetype	Yes	Yes
Complex linetype	Yes	Yes, if *ltypeshx.shx* found
Lineweight	Yes	Yes
Database links	**No**	**No**

Attachment	Open	Import
DGN	**No**	**No**
DWF	**No**	**No**
Image	Yes	Yes, if found
OLE	Yes	Yes
Sheetset	**No**	**No**
XRef	Yes	Yes

AutoCAD Named Objects	Transferred to Inventor
Blocks	Converted to symbols; attributes converted to text
Color books	**No**: Inventor lacks color books
Dimension styles	Names and properties
Gradients	**No**: Inventor lacks gradients
Hatch patterns	**No**: patterns hardcoded in Inventor
Layouts	Names only
Layers	Names and properties
Linetypes	**No**, but Inventor reads *.lin* files
Lineweights	**No**: hardcoded in AutoCAD and Inventor
Materials	**No**
Plot styles	**No**: Inventor lacks plot styles
Shapes	**No**
Text styles	Names and properties
Multiline styles	**No**: Inventor lacks multilines
Table styles	**No**
Viewports	Paper space only; names, positions, and scales
Visual styles	**No**: visual styles hardcoded in Inventor

Exporting Inventor Environments

The following table shows which of the Inventor environments can be exported to AutoCAD:

Inventor Environment	File Extension for Inventor	Save Copy As for AutoCAD
Part	IPT	SAT
Assembly	IAM	SAT
Inventor drawing	DWG, IDW	Inventor Drawing (DWG), AutoCAD Drawing (DWG), DXF
Presentation	IPN	None available
AutoCAD drawing	DWG*	Inventor Drawing (DWG), AutoCAD Drawing (DWG), DXF
Sheetmetal flat patterns	IPT**	SAT
Sketches, planar faces	...	***

Notes:

*) The Save As command can be used only the first time a .dwg template is saved to an .idw file and the first save of an existing .dwg. Once saved, the Save As or Save Copy As commands can be used to save to the other file type.

***) Right-click the flat pattern in model browser, and then choose **Save Copy As** SAT, DWG, or DXF.*

****) Right-click the sketch in model browser, and **Save Copy As** DWG or DXF. DWF is compatible back to AutoCAD 2000; DXF is compatible back to AutoCAD Release 12.*

Castle Learning Resource Centre

Part IV

Appendices

Appendix A

AutoCAD & Inventor Jargon

As someone proficient in AutoCAD, you are familiar with many terms and concepts specific to CAD. You know how to draw lines and circles, how to save and print drawings. But, having being developed independently of AutoCAD, Inventor uses some jargon that differs from AutoCAD.

The following tables cross-reference terms and command names between the two CAD programs. But note that many of the "equivalents" are not truly identical, but are the closest match. For example, an AutoCAD object snap is only effective for the split-second that it takes AutoCAD to find the snapped location, whereas Inventor's equivalent — constraints — are sticky and continue to remember the required connection.

In addition, both programs have commands, terms, and functionality for which the other has no equivalent or even near-equivalent. For example, Inventor has no equivalent for AutoCAD's plot styles, and AutoCAD has nothing even close to Inventor's parametric functionality.

AutoCAD Jargon

When you know the term in AutoCAD, but wonder what the equivalent might be in Inventor, use the following table for finding cross references.

AutoCAD Term	Equivalent in Inventor
Array	Pattern
Background	Scene
Block	Component
Break	Split
Close	Return
DesignCenter	ContentCenter
Dimension	General Dimension
Dist	Measure
Drawing	Document

AutoCAD Term	Equivalent in Inventor
eTransmit	Pack and Go
Erase	Delete
Fields	Property expression
Helix	Coil
Layout	Sheet
LightList	Lighting styles
Materials	Surface texture
MLeader	Balloon
MView	Drawing View
Object Snap	Constraint
Options	Application options, Document options
Orbit	Rotate
Ortho mode	Horizontal and vertical constraints
Publish	Multisheet plot
Perspective	View mode
Plan	Fit to
Plot	Print
Polar array	Circular pattern
Publish	Batch publish
Section	Slice
Slice	Split
SolView	Section view
Style	Format text
Tolerance	Feature control frame
UCS	Work plane
Viewports	Drawing views
Visual styles	Display mode
WalkFlySettings	Animation parameters
Workspace	Environment
Xline	Work axis
Zoom object	Zoom select

When you come across other unfamiliar terms, add them below:

AutoCAD Term	Equivalent in Inventor
_____	_____
_____	_____
_____	_____
_____	_____
_____	_____

Inventor Jargon

When you come across an Inventor term that puzzles you, use the following table to find the equivalent term or command in AutoCAD.

Inventor Term	Equivalent in AutoCAD
Animation parameters	WalkFlySettings
Application options	Options
Batch publish	Publish
Balloon	MLeader
ContentCenter	DesignCenter
Circular pattern	Polar array*
Coil	Helix
Constraint	Object snap
Component	Block
Delete	Erase
Display mode	Visual styles
Document	Drawing
Document options	Options
Drawing view	MView
Environment	Workspace
Feature	Extrusion, revolve, and so on
Feature control frame	Tolerance
Fit to	Plan
Format text	Style
General dimension	Dimension
Lighting styles	LightList
Measure	Dist
Multisheet plot	Publish
Pack and Go	eTransmit
Pattern	Array
Print	Plot
Property Expression	Fields
Return	Close
Rotate	Orbit
Scene	Background
Section view	SolView
Sheet	Layout
Slice	Section
Split	Slice or beak
Surface texture	Materials

Note:
**) AutoCAD used to call this "circular" arrays, but changed to "polar" many years ago.*

continued

Inventor Term	Equivalent in AutoCAD
View mode	Perspective
Work axis	Xline
Work plane	Temporary UCS
Zoom select	Zoom object

When you come across more unfamiliar terms, add them below:

Inventor Term	Equivalent in AutoCAD
_____	_____
_____	_____
_____	_____
_____	_____
_____	_____
_____	_____
_____	_____

No Equivalent

The following are some terms and concepts that have no equivalent in the other CAD program:

AutoCAD

Aerial View

Clean Screen

Draw order

MLine

Named Views

Plot style

Redraw

Regen

Revision Cloud

Sheet Set manager

Viewports

Inventor

Associativity

Design accelerators

Driven constraints

iMates

iParts

Kinematics

Motion constraints

Weldments

Appendix B

Inventor & AutoCAD:
Import & Export Formats

Inventor supports some of the same files as does AutoCAD. The tables in this appendix show you at a glance the file formats handled by both programs, and which they have in common.

Formats Native to Inventor

Inventor and AutoCAD directly read the following file formats; the nearest AutoCAD equivalents are listed below.

Type	Inventor	Equivalent in AutoCAD
Drawing	IDW, DWG*	DWG*
Features	IDE	...
Assembly	IAM	DST**
Presentation	IPN	...
Board	IDB	...
Projects	IPJ	DST**
Templates	IPT	DWT
Standards	STYXML	DWS
Linetypes	LIN	LIN
Hatches	...	PAT
Fonts	TTF	SHX, TTF

Notes:
*) Inventor's DWG format adds custom ObjectARX objects to AutoCAD's; both CAD packages can read each other's .dwg files, although AutoCAD requires the TrueConnect object enabler.
**) Sheetsets are the closest AutoCAD has to assemblies and projects.

Supported Import Formats

Inventor and AutoCAD are able to import the following file formats. Those listed as "in beta" are available as a download from labs.autodesk.com.

Type	Inventor	AutoCAD
DWG	AutoCAD DWG	Inventor DWG
Interchange	DXF	DXF
ASCII ACIS	SAT	SAT
IGES	IGS, IGE, IGES	...
STEP	STP, STE, STEP	...
Pro/Engineer	PRT, ASM ...	
MicroStation	...	DGN
SolidWorks	*(in beta)*	...
Parasolid *(in beta)*	X_B, X_T	...
UGS NX	*(In beta)*	...
PTC Granite *(in beta)*	G, NEU	...
Pro/Engineer	*(in beta)*	...

Supported Export Formats

Inventor and AutoCAD programs export drawings in the following file formats. Those listed as "in beta" are available as a download from labs.autodesk.com.

Type	Inventor	AutoCAD
DWG	AutoCAD DWG	DWG*
ePublishing	2D *and* 3D DWF	2D *and* 3D DWF
Raster	BMP, GIF, *etc.*	BMP, GIF, *etc.*
IGES	IGS, IGE, IGES	...
JT	JT	...
SAT	SAT	SAT
STEP	STP, STE, STEP	...
MicroStation	...	DGN
Parasolid *(in beta)*	X_B, X_T	...
PTC Granite *(in beta)*	G, NEU	...
OpenGL**	XGL	...
Compressed OpenGL**	ZG	...

Notes:
*) *AutoCAD can save Inventor-created DWG files.*
**) *Used by Autodesk Streamline, but superceeded by DWF.*

Index

Notes